Symmetry and Stereochemistry

J. D. DONALDSON
B.Sc., Ph.D. (Aberdeen), D.Sc. (London), A.R.I.C.

and

S. D. ROSS
B.Sc., M.Sc., Ph.D. (London), A.R.I.C., M.Inst.P.

Halsted Press Division
John Wiley & Sons, Inc.
New York

Published in the United States and Canada by Halsted Press Division,
John Wiley & Sons, Inc., New York

First published 1972

ISBN 0 470-21778-2
Library of Congress
catalog card no. 72-2296

Printed by photolithography and made in Great Britain at
The Pitman Press, Bath

Preface

This book is based on the experiences of a twenty-five lecture course given to second year undergraduate students. The main aim of both book and course is to provide a unified approach to the concepts and use of symmetry. There are some excellent texts containing treatments of the subject designed specifically to cover certain applications, notably X-ray crystallography and vibrational spectroscopy, but few attempts have been made to provide a unified approach in a student text.

We begin by considering the ways in which an object can be repeated in space by translation, rotation, reflection and inversion. Molecules can be discussed in terms of symmetry which involves only rotation, reflection and inversion and of combinations of these symmetry elements. This aspect of the topic is developed in a Chapter on point group symmetry. Crystals can only be described in terms of all the symmetry elements and their combinations and this theme is developed under the heading space-group symmetry. We have introduced the mathematical concepts of group theory as they are required to enable us to expand the usefulness of concepts of symmetry and to provide a basis for discussions on and systematisation of the various physical properties of molecules and crystals. The last two chapters of this book deal with applications of the concepts discussed previously to problems of interest to the physical scientist.

Throughout this book we have attempted to illustrate both the development and the application of the concept of symmetry by means of numerous examples. The reader is given the opportunity to follow through worked examples both in the main text and in the questions at the ends of the chapters. We believe that treatment of symmetry in this way leads to the most satisfactory presentation of the subject for undergraduates.

Finally, we should like to thank all those who assisted in the production of this book; this includes those who prepared the typescript and those of our colleagues and students with whom we have had many valuable discussions on the topics dealt with in the text.

J.D.D.
S.D.R.

Contents

CONTENTS

1

Introduction

The concept of symmetry has applications in many aspects of chemistry and the use of a knowledge of symmetry and of the mathematical principles governing it is often a common feature in otherwise unrelated work. For example the determination of the crystal structure of a protein, the interpretation of the vibrational spectrum of a vapour, and a molecular-orbital calculation on a conjugated organic molecule are related by a common requirement to make use of arguments based on symmetry. The importance of a knowledge of the basic symmetry elements, of their combinations, of the mathematical theory which describes their operation, and of the applications of symmetry in various aspects of chemistry cannot be overstressed.

External Symmetry in Crystals

The external faces of crystals generally provide a chemist with his first introduction to the importance of symmetry. The external faces of a crystal are of course related to the internal symmetry of the material and represent the appearance at the surface of the slowest-growing of the many possible faces of the crystal. The element copper has a face-centred cubic structure (Figure 1.1) and crystallises in the form of an octahedron (Figure 1.2). The faces of the octahedron represent the slowest growing planes of the copper crystal, which are the planes containing all of the atoms in the square diagonals of three of the cube faces. This is illustrated by the plane through atoms 1 to 6 in Figure 1.3. Growth along this plane produces the face labelled A in the octahedron in Figure 1.2.

The external form of the crystals of a material is not invariant because any factor which affects the rate of growth of the crystal in various directions will affect the types of faces appearing at the surface. Alums crystallise in a form based on an octahedron with extra faces (Figure 1.4) but, if the crystals are grown at the bottom of a beaker, no growth can occur on the face which

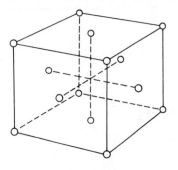

Figure 1.1   Face-centred cubic structure of copper

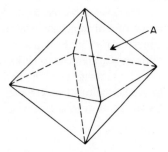

Figure 1.2   Octahedral external form of copper crystals

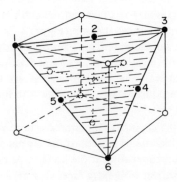

Figure 1.3   Plane of growth of copper crystal which forms one of the octahedral faces

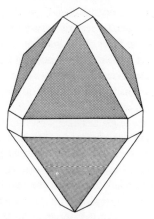

Figure 1.4   External structure of an alum crystal

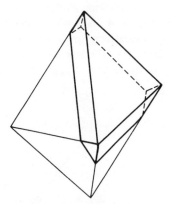

Figure 1.5   Triangular tablet form of an alum crystal formed when growth along one
                  face is prevented

lies on the beaker and a triangular tablet (Figure 1.5) is formed instead of a
complete octahedron.

   Sodium chloride crystallises as cubes from aqueous solution but addition of
a small quantity of urea to the solution results in the development of small octahedral faces at the corners of the cube and addition of about 10 per cent of urea
results in the crystallisation of the sodium chloride as octahedra. The urea does
not enter the sodium chloride lattice but, because it is preferentially absorbed on
planes parallel to the octahedral faces, it suppresses the growth of the crystal along
these planes and causes them to develop slowly enough to appear at the surface.
The external shape of a sodium chlorate crystal can likewise be changed from its
normal cubic form when grown from aqueous solution to a tetragonal form by
the addition of about 8% of sodium borate.

Although the final shape of a crystal does depend upon the conditions under which it grows, two of its properties remain invariant; (i) its internal structure, and (ii) the angles between the faces at the surface. The internal structure of sodium chloride, represented by the relative positions of the atoms in the crystal, remains the same whatever the external form and this can be demonstrated by a constancy in the X-ray diffraction patterns for the various shapes of crystal. If the faces which appear at the surface can be identified, it can readily be shown that the angle between any two faces remains constant irrespective of any changes in the relative surface areas occupied by these faces. The external faces of a crystal are generally identified by a set of three integers (*h k l*) called *Miller Indices*. These indices are derived from the intercepts that the plane forming the external face makes on the major axes of the crystal. If a crystal has major axes x, y, z of length a, b, c respectively as in Figure 1.6 then the Miller indices are derived from the intercepts (in fractions of the axes lengths a, b and c) of the plane with these axes. The indices are obtained by taking the reciprocals of the intercepts and clearing the fractions, if necessary, by multiplying through by a small integer. A plane which cuts the three major axes with intercepts of $\frac{2}{3}a$, $\frac{1}{3}b$ and $\frac{1}{2}c$ respectively would have Miller indices (3 6 4), i.e. formed by taking the reciprocals of the intercepts, 3/2, 3/1 and 2/1 and clearing the fractions by multiplying through by 2 to give (3 6 4). Miller indices of

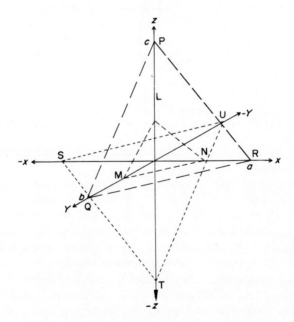

Figure 1.6   Diagram to show planes with Miller indices (2 2 3), (1 1 1) and (1 1 1)

planes forming crystal faces are conventionally placed in brackets as shown. If a
plane cuts any of the major axes with a negative intercept a negative sign is placed
above its Miller index. For example a plane which cuts the three major axes
with intercepts of $-\frac{2}{3}a$, $-\frac{1}{3}b$ and $\frac{1}{2}c$ respectively will have the Miller indices
$(\bar{3}\ \bar{6}\ 4)$. The intercept of a plane on an axis with which it is parallel is infinity
and its Miller index is 0. In Figure 1.6 the plane LMN has intercepts of $\frac{1}{2}a$, $\frac{1}{2}b$
and $\frac{1}{3}c$ and thus has Miller indices (2 2 3). The plane PQR has intercepts of
$1a$, $1b$ and $1c$ along x, y and z and is thus the (1 1 1) face and the plane STU
with intercepts $-1a$, $-1b$ and $-1c$ is the $(\bar{1}\ \bar{1}\ \bar{1})$ face. The faces of the octahedron (Figure 1.2) are planes of the type (1 1 1), $(\bar{1}\ 1\ 1)$, $(\bar{1}\ \bar{1}\ 1)$ etc. and
the faces on the cubic crystals of sodium chloride are of the type (1 0 0),
(0 1 0), $(0\ 0\ \bar{1})$ etc.

Stereographic Projections

One of the most convenient ways of representing the fact that angles between
faces are constant for a given crystalline material is by means of a stereographic
projection. This is a projection of the external symmetry of the crystal on the
equatorial circle of a sphere which has the centre of the crystal at its centre. These
projections of crystal faces are particularly useful in geological classifications of
crystals but, in general, stereographic projections are a convenient means of
representing three dimensional symmetry in two dimensions. The construction
of a stereographic projection will be illustrated in this chapter for an octahedral
crystal, but more general uses of these projections will be discussed in later
sections.

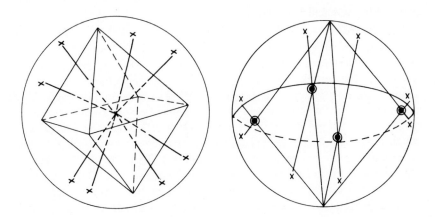

Figure 1.7
        and   Construction of the stereographic projection of an octahedral crystal
Figure 1.8

The stereographic projection of an octahedral crystal is constructed by placing the centre of the crystal at the centre of a sphere. The sphere's radii which are perpendicular to the faces of the octahedron are then drawn as shown in Figure 1.7 and the points at which these radii cut the surface of the sphere are marked. The stereographic projection is then obtained by joining all of the points marked in the northern hemisphere to the south pole of the sphere and all points in the southern hemisphere to the north pole as shown in Figure 1.8. The intersections that lines from the north pole make with the equatorial plane of the sphere are marked by open circles (○) and the intersections of lines from the south pole are marked by closed circles (●). For the octahedral crystal, lines from both poles cut the equatorial plane at the same point and this is represented by a closed circle within an open circle (⊙). The stereographic projection for the octahedral crystal consists of the equatorial circle with the points of intersection of the lines from the poles marked as in Figure 1.9.

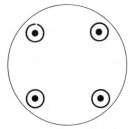

Figure 1.9   The stereographic projection of an octahedral crystal

Internal Structure and Symmetry

Tartaric acid is an optically active material and the crystals of its *dextro-* and *laevo-* forms are mirror images of each other as shown in Figure 1.10. The molecules of tartaric acid and the atoms in the molecules are distributed in space within the crystals in strict accordance with the laws of symmetry. In fact, in spite of the large variety of chemical substances which are crystalline solids and in spite of the possible variations in external symmetry, any crystal must belong to one of only 230 possible crystal symmetry types. These are the *230 space groups*. If we extract from the crystal one of the molecules of tartaric acid then the symmetry of that molecule in isolation is characterised by its *point group.* Most molecules (including tartaric acid) belong to one of only 32 crystal point groups. There are some molecules, however, which do not belong to one of the 32 crystal point groups. The 32 groups arise because there are restrictions on the axial symmetry of close-packed three-dimensional solids. Thus 5-, 7-, 8-, and infinite-fold axes (see page 73) are permitted in molecules but not in crystals.

Many of the properties of the molecule of tartaric acid are governed by its

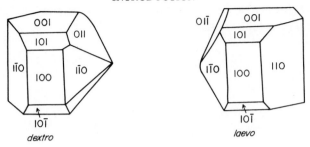

Figure 1.10   The external form of the *dextro-* and *laevo-* forms of tartaric acid

symmetry. For example its optical activity and vibrational spectrum are governed by, and can be described in terms of, its symmetry. The molecular orbital description of the bonding in the molecule is also based on the symmetry properties of the molecular orbitals and of the atomic orbitals from which they are formed.

In the course of this book we shall develop from a consideration of the basic symmetry elements and of their combinations to a discussion of space and point group symmetry. We shall introduce the mathematical concepts of group theory as they are required to enable us to expand the usefulness of concepts of symmetry and to provide a basis for discussions on and systematisation of the various physical properties of molecules and crystals.

PROBLEMS

1. What will be the final shapes of two-dimensional crystals of shapes A and B if the slowest growing faces are those indicated by the arrows?

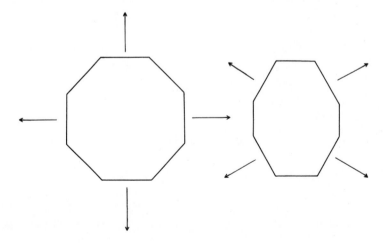

2. What are the Miller indices of the planes which have the following intercepts along the major axes of a crystal:

   (i) $a, \frac{1}{2}b, \frac{1}{2}c$

   (ii) $\frac{3}{4}a, b, \frac{1}{2}c$

   (iii) $\frac{1}{2}a, b, -c$

   (iv) $\frac{1}{4}a, -\frac{1}{2}b, -c$?

3. Draw a diagram to represent the planes with Miller indices (3 2 2) and (3 2 $\bar{2}$) in a crystal system with three axes at right angles and of length a, b and c respectively.

4. What are the Miller indices of the faces of a regular tetrahedron?

2

Symmetry Elements and Operations

Symmetry Elements

There are only four ways in which an asymmetric object can be repeated in space. These are the four basic symmetry elements of:

Translation — the three dimensional repeat pattern
Reflection — symmetry about a plane
Rotation — symmetry about an axis
Inversion — symmetry about a point

These symmetry elements will be described in detail using the figure 7 as an asymmetric object. There are two distinct systems of nomenclature in use: that used by crystallographers (the Hermann-Mauguin system) and that used by spectroscopists (the Schoenflies system). Both systems of nomenclature are introduced at this stage to enable the reader to become conversant with the two and to be able to interconvert them.

TRANSLATION

In translation symmetry the asymmetric object is repeated in space at regular intervals by moving it in a given direction through a constant distance t. For example, translation of the figure 7 along a line by a translation t gives the one dimensional repeat pattern

$$7 \leftarrow t \rightarrow 7 \leftarrow t \rightarrow 7$$

If the object is repeated in two dimensions by two translations (t_1 and t_2) the repeat pattern is that of the plane array of 7s shown in Figure 2.1 and if the object is repeated in three dimensions by three translations (t_1, t_2 and t_3) the repeat pattern is that of the space array of 7s shown in Figure 2.2. For example the iron atoms in the body-centred cubic structure of the α-modification of the element

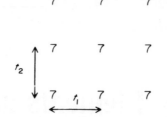

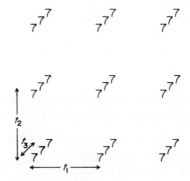

Figure 2.1   A plane array of asymmetric objects

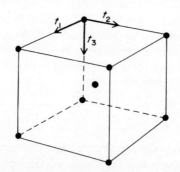

Figure 2.2   A space array of asymmetric objects

Figure 2.3   Cubic space array of iron atoms in α-Fe showing the three mutually
                  perpendicular unit translations

provide an example of a space array of atoms repeated in space by three equal
translations at right angles (Figure 2.3).

REFLECTION

In reflection symmetry the asymmetric object is repeated in space by reflecting
it across a mirror plane.

The symbols for reflection are m (crystallographic) and σ (spectroscopic). Figure 2.4 shows a number of examples of molecules containing mirror planes. The Hermann-Mauguin system of nomenclature does not distinguish between the various orientations of a mirror plane in a molecule, but the Schoenflies system recognises three types of plane. These are designated σ_h, σ_v and σ_d. The subscripts stand for horizontal, vertical and diagonal respectively, and refer to the orientation of the plane with respect to the principal rotation axis (see next section) which conventionally is taken as being vertical. The need to refer mirror planes to a rotation axis does not arise in crystals because their orientation can instead be referred to the major crystal axes. Some of the molecules in Figure 2.4

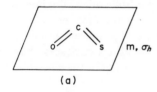

(a)

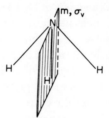

One side is vertical mirror
image of the other

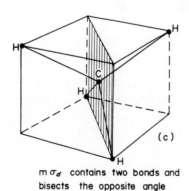

(c)

m σ_d contains two bonds and
bisects the opposite angle

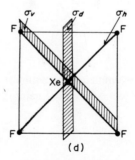

(d)

Plane of molecule is σ_h
Planes containing two bonds are σ_v
Planes bisecting bond angles are σ_d

Figure 2.4   Types of mirror planes which occur in molecules

contain more than one mirror plane of a given type but only one example of each type is shown for any molecule.

ROTATION

In rotation symmetry a right-handed asymmetric object is repeated in space by rotating it through an angle of $360°/n$, where n is an integer or infinity, to give another right-handed object. Figure 2.5 shows how a 7 can be repeated by rotations through $360°/n$ to give n-fold axes where n has any integer value from one to seven. The symbols for rotation axes are n (crystallographic) and C_n (spectroscopic). Examples of molecules containing each of the n-fold axes are also shown in Figure 2.5. The molecules used to describe the various axes are: 1-fold, HC1; 2-fold, formaldehyde; 3-fold, chloromethane; 4-fold, xenon oxy-fluoride; 5-fold, ruthenocene; 6-fold, benzene; 7-fold, the tropylium ion. The axes are not the only symmetry elements present in the molecules in Figure 2.5. All of them, for example, contain at least one mirror plane. Some of them contain other rotation symmetry axes, benzene has 2-fold and 3-fold axes in addition to the 6-fold axis, xenon oxyfluoride has 2-fold axes in addition to the 4-fold axis and the molecular axis of HC1 is an axis of infinite order. The internuclear axis of any linear molecule must, in fact, be an axis of infinite order because any rotation, however small, about this axis leaves the molecule in a configuration indistinguishable from the original. If a molecule contains axes of different orders then the axis of highest order is the principal axis of the molecule. Thus the 6-fold axis of benzene is its principal axis.

It is convenient, for the purposes of producing the projections described in later chapters, to have diagrammatic symbols to represent the rotation axes. The symbols used for rotation axes are the closed polygons shown on the axes in Figure 2.5. For example the symbols for 2-, 3-, 4- and 6-fold axes are

INVERSION

In inversion symmetry an asymmetric object with coordinates (x,y,z) is converted to an object with co-ordinates $(-x,-y,-z)$ by inversion through a point.

7

∠

The symbols for inversion (centre of symmetry) are $\bar{1}$ (crystallographic) and i (spectroscopic). In a molecule the centre of inversion may or may not be occupied by an atom. Figure 2.6 shows the positions of the centres of symmetry in the molecules XeF_4, with centre at the Xe atom, and C_2H_2, with centre in the middle of the C≡C bond. Inversion can be regarded as a special case of reflection, i.e. reflection through a point rather than through a plane.

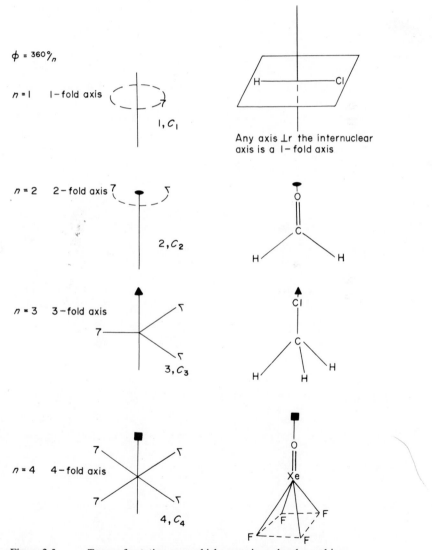

Figure 2.5   Types of rotation axes which occur in molecules and ions

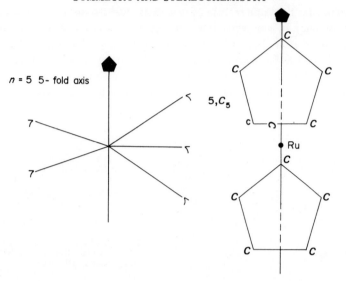

$n = 5$ 5- fold axis

$5, C_5$

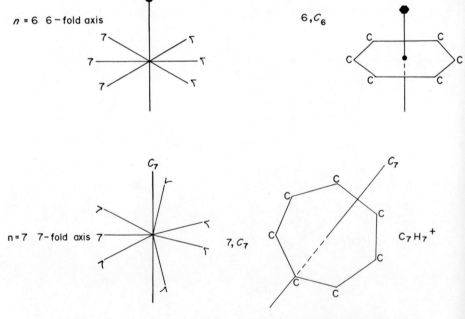

$n = 6$ 6 – fold axis

$6, C_6$

C_7

C_7

n = 7   7- fold axis

$7, C_7$

$C_7 H_7{}^+$

Figure 2.5 (cont.)   Types of rotation axes which occur in molecules and ions

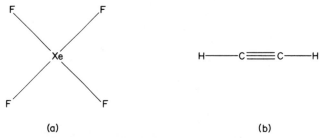

(a)   (b)

Figure 2.6   Typical locations of centres of symmetry (inversion) in molecules
                    (a) centre of symmetry is occupied by Xe atom
                    (b) centre of symmetry is midway along the $C \equiv C$ bond

ROTATION, REFLECTION AND INVERSION SYMMETRY IN CRYSTALS

We have used molecules and ions to illustrate rotation, reflection and inversion
symmetry, but these elements can also be found in crystalline materials. For
example the following symmetry elements can be picked out in the cubic unit
cell of sodium chloride as illustrated in Figure 2.7

(i) 2-fold axes, parallel to a square diagonal, through any sodium or chloride
ion

(ii) 3-fold axes, parallel to the cube diagonal, through any sodium or
chloride ion

(iii) 4-fold axes, parallel to a cube edge, through any sodium or chloride
ion

(iv) mirror planes — all planes which contain two of the cube edges are
mirror planes

(v) centres of symmetry at the centres of any sodium or chloride ion.

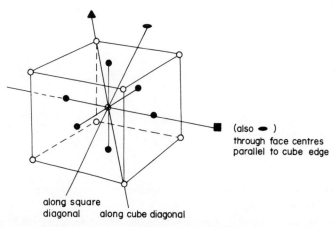

Figure 2.7 (a)   Typical rotation symmetry axes in crystals of sodium chloride

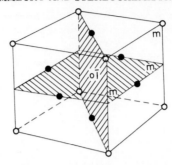

Figure 2.7(b)   Centre of symmetry and typical mirror planes in sodium chloride

Symmetry Operations

We have so far considered symmetry in terms of the generation of an array of asymmetric objects in space. For example a 3-fold rotation axis is an example of a symmetry element; it reproduces an asymmetric object in space to give the pattern of Figure 2.8a. A symmetry operation, however, is the process by which an array equivalent to the original can be produced by operation of a symmetry element. For example, if we start with an array of 7s in the orientation of Figure 2.8b in which 7A is in position 1, 7B is in position 2 and 7C is in position 3, and then perform the operation of rotation through 360/3 = 120° the array takes up the configuration of Figure 2.8c with 7A in position 2, 7B in position 3 and 7C in position 1. This new configuration is equivalent to but not identical

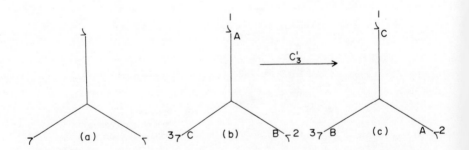

Figure 2.8(a) An array of asymmetric objects generated by a 3-fold rotation axis

(b) The original and

(c) the new orientation of the array of asymmetric objects produced by a rotation through 120° about a 3-fold axis with the original. The difference between a symmetry element and a symmetry operation is that the operation permits consideration of the orientation of a molecule or an array while the element considers only the repetition of the asymmetric unit of the array in space.

If we are considering new orientations produced by symmetry operations, translation symmetry is not relevant because it produces a displacement in space but not a new orientation. We need only consider reflection, rotation and inversion and for this purpose it is convenient to discuss molecules as examples.

The reflection symmetry element is the plane of symmetry, m or σ, which divides the molecule into two halves which are mirror images of each other. The rotation symmetry element is the axis of symmetry, n or C_n; when a molecule is rotated by $360°/n$ about such an axis it assumes a configuration equivalent to the original.

The inversion symmetry element is the centre of symmetry, $\bar{1}$ or i. If the centre of symmetry is placed at the origin of a Cartesian co-ordinate system, then to any atom at (x,y,z) there corresponds an identical atom at $(-x,-y,-z)$.

The process, by which a configuration equivalent to the original is produced, is the symmetry operation. Reflection at a mirror plane can produce only one such configuration, and the reflection element is said to generate one symmetry operation. Both the operation and the element are designated by the same symbol. Figure 2.9 shows that each reflection at a mirror plane in BF_3 produces one configuration equivalent to the original.

The process of rotation about an n-fold axis, however, produces n such configurations, and the rotation element is said to generate n symmetry operations.

When a configuration equivalent to the original is produced by a rotation through

$$\frac{k}{n} \times 360°$$

about an axis, the symmetry operation is designated C_n^k, k can have any integer value from 1 to n. For the C_3 axis in BF_3, k takes the values 1, 2, 3 and, in general, for any C_n axis, n equivalent configurations are produced. When $k = n$, (as for C_3^3 in Figure 2.10) the configuration is not merely equivalent to, but identical with, the original and the operation C_n^n is thus equivalent to the so-called identity operation I. Many texts designate this as E, but we have preferred I so as to avoid confusion with another conventional meaning for E. If we look at Figures 2.9 and 2.10 we can also see that the performance of two successive symmetry operations leads to a configuration equivalent to the original. Now it is always true that there is some single operation by which this new configuration could have been attained. Consider the initial configuration, which we can represent as A1B2C3 (i.e. fluorine atom A occupies position 1, and so on). If we perform the operation σ_{v1}, which is reflection at the mirror plane going

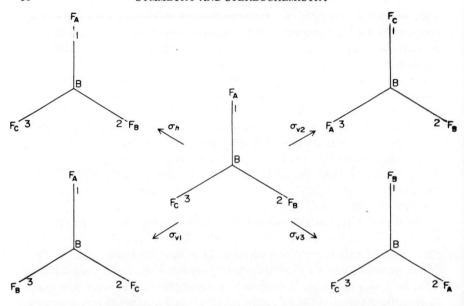

Figure 2.9   Equivalent configurations produced by reflection at mirror planes in the BF₃ molecule

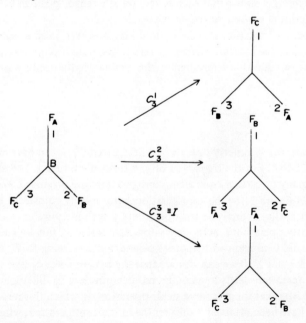

Figure 2.10   The three equivalent configurations generated by the 3-fold axis of the BF₃ molecule

through position 1, we get A1B3C2. Now perform σ_{v2}, which is reflection at the plane going through position 2. The atom now in position 2 (atom C) is unaffected; A and B change places, giving the configuration A3B1C2. Figure 2.10 shows that this configuration could have been attained in one step, by the operation C_3^2. Thus we can write, symbolically,

$$\sigma_{v2}\,\sigma_{v1} \equiv C_3^2$$

which means that if we perform first σ_{v1} and then σ_{v2}, the effect is the same as if we had performed C_3^2 only. The first operation to be performed is always written to the right of the second, because, in general, the order of performance makes a difference. If we had performed σ_{v2}, followed by σ_{v1}, we would get

$$A1B2C3 \xrightarrow{\ \sigma_{v2}\ } A3B2C1 \xrightarrow{\ \sigma_{v1}\ } A2B3C1,$$

and this last configuration would be achieved by the single operation C_3^1,

$$\text{i.e.}\quad \sigma_{v1}\,\sigma_{v2} \equiv C_3^1.$$

MULTIPLICATION TABLES

For any molecule, we can construct a multiplication table of symmetry operations, showing the effect, or product, of any two successive operations.

Consider N_2F_2 in the *trans* form. This has a centre of symmetry, at the mid-point of the N–N bond; the plane of the molecule is a plane of symmetry, and since it is perpendicular to the only axis of rotation, it is a σ_h; the axis itself is a C_2 and thus generates the two operations $C_2^1, C_2^2 (\equiv I)$.

Table 2.1. Multiplication table of symmetry operations for *trans*-N_2F_2

		I	C_2^1	σ_h	i	first operation
second	I	I	C_2^1	σ_h	i	
operation	C_2^1	C_2^1	I	i	σ_h	
	σ_h	σ_h	i	I	C_2^1	
	i	i	σ_h	C_2^1	I	

It is important to notice the following points about Table 2.1:

(i) The product of any two symmetry operations of the molecule is itself a symmetry operation of the molecule;

(ii) Whichever operation we carry out first, there is always a second operation such that the product of the two is the identity. This second operation is called the inverse of the first.

SYMMETRY AND STEREOCHEMISTRY

Now consider the ammonia molecule. This has six symmetry operations, three of which, $C_3^1, C_3^2, C_3^3 (\equiv I)$ are generated by the C_3 axis. The remaining three are each generated by a plane of symmetry. Each plane of symmetry contains one of the N–H bonds, and the planes are σ_v since the C_3 axis lies in each of them (and is in fact the line formed by their intersection). If we label the planes as in Figure 2.11, we obtain the multiplication table of Table 2.2.

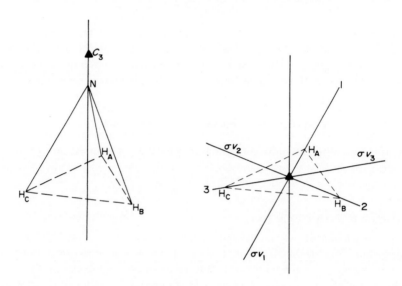

Figure 2.11   Symmetry elements of the NH_3 molecule (mirror planes shown in plan)
Note that the numbering of the σ_v is with reference to fixed positions; if atom A is originally in position 1, the plane passing through that position is always σ_{v_1} irrespective of which atom occupies position 1

Table 2.2. Multiplication table of symmetry operations for NH_3

		I	C_3^1	C_3^2	σ_{v_1}	σ_{v_2}	σ_{v_3} first operation
	I	I	C_3^1	C_3^2	σ_{v_1}	σ_{v_2}	σ_{v_3}
second operation	C_3^1	C_3^1	C_3^2	I	σ_{v_3}	σ_{v_1}	σ_{v_2}
	C_3^2	C_3^2	I	C_3^1	σ_{v_2}	σ_{v_3}	σ_{v_1}
	σ_{v_1}	σ_{v_1}	σ_{v_2}	σ_{v_3}	I	C_3^1	C_3^2
	σ_{v_2}	σ_{v_2}	σ_{v_3}	σ_{v_1}	C_3^2	I	C_3^1
	σ_{v_3}	σ_{v_3}	σ_{v_1}	σ_{v_2}	C_3^1	C_3^2	I

The statements (i) and (ii) made above about Table 2.1 are also true of Table 2.2. However, there is one important difference between the two tables. In Table 2.1, it is immaterial which operation is carried out first; if we interchange the rows and columns, the table remains unaltered. Further, according to the definition of an inverse given in (ii), every operation is its own inverse. Neither of these statements is true of Table 2.2; it does in most cases make a difference which operation is carried out first, and not every operation is its own inverse. In these respects, molecules whose highest order rotation axis is 2 always have a multiplication table like Table 2.1; those whose highest order rotation axis is greater than 2 have a table like Table 2.2.

The following general rules are helpful in the construction of multiplication tables:

1. The product of two rotations is always a rotation.
2. The product of two reflections is always a rotation. If the two reflection planes intersect at an angle ϕ, the resulting rotation is through 2ϕ.
3. The product of a rotation and reflection in a plane containing the rotation axis is a reflection in another plane containing the axis. If the angle of rotation is ϕ, the angle between the two reflection planes is $\phi/2$.
4. The following operations *commute* (i.e. the order in which they are performed is immaterial):
    (a) two rotations about the same axis;
    (b) two reflections in perpendicular planes;
    (c) two rotations by $180°$ about perpendicular axes;
    (d) a rotation and a reflection in a plane perpendicular to the rotation axis;
    (e) any operation and the inversion operation;
    (f) any operation and the identity.
5. Every operation occurs once in each row and column of the multiplication table.

In this chapter we have discussed the basic symmetry elements of translation, rotation, reflection and inversion and have considered their operations. Molecules can be discussed in terms of symmetry which involves only rotation, reflection and inversion and of combinations of these symmetry elements. This aspect of symmetry and stereochemistry is developed in Chapter 3. Crystals can only be described in terms of all of the symmetry elements (including translation) and their combinations, and this theme is developed in Chapter 4. The multiplication tables constructed in this chapter contain the necessary information which makes it possible to apply symmetry considerations to the descriptions of physical properties. This information, however, is not in a form in which it can be readily used. The application of group theory described in Chapter 5 depends on the conversion of this information into a usable form.

PROBLEMS

1. In octahedral crystals of NaCl, which are the directions of the 4-fold and 2-fold axes?

2. Pick out the mirror planes in the following molecules or ions:

   (a) HOCl   (d) XeF_4

   (b) H_2O (e) CCl_4

   (c) NO_3^- (f) SF_6

3. Pick out the rotation axes in the following molecules or ions:

   (a) N_2O (d) cyclopentadienyl anion

   (b) CO_2 (e) $PtCl_4^{2-}$

   (c) SO_2 (f) ethane (staggered configuration)

4. Which of the following molecules or ions contains a centre of symmetry?

   (a) N_2 (e) $SiCl_4$

   (b) $AuCl_4^-$ (f) naphthalene

   (c) benzene   (g) cyclohexane (chair)

   (d) SF_6 (h) cyclohexane (boat)

5. Construct multiplication tables for the symmetry operations of:

   (a) formaldehyde

   (b) $XeOF_4$

   (c) *p*-dichlorobenzene

6. Verify the general rules 1–5 (page 21) in so far as they apply to the tables constructed in the previous problem.

3

Point Group Symmetry

Molecular symmetry is characterised by three of the four basic types of symmetry element, these three being reflection, rotation and inversion. We can distinguish two types of symmetry element, simple and compound. The simple elements of reflection, rotation and inversion have been described in Chapter 2, but we can also form compound symmetry elements by combining rotation with either reflection or inversion to give rotor-reflection and rotor-inversion axes respectively.

ROTOR-REFLECTION AXES

In these compound elements, an asymmetric object is rotated through $360°/n$ and then reflected in a mirror plane. Such axes are designated $\tilde{n}$ in crystallographic and S_n in spectroscopic notation. Figure 3.1 shows the 1-fold and 2-fold rotor-reflection axes.

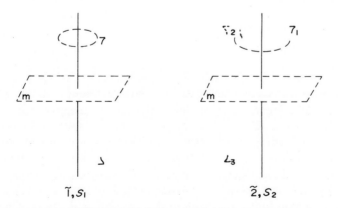

Figure 3.1   1-fold and 2-fold rotor-reflection axes

The 1-fold rotor-reflection axis requires the asymmetric object (a figure 7) to be rotated through 360°/1, that is, back to its original position, and then reflected in the mirror plane. The 2-fold rotor-reflection axis requires a rotation from position 7_1 through 360°/2 to position 7_2, followed by reflection in a mirror plane to position 7_3. (7_1 and 7_3 are thus related to one another by a 2-fold rotor-reflection axis.)

ROTOR-INVERSION AXES

In these compound elements, an asymmetric object is rotated through 360°/n and then inverted through a point. Such axes are designated $\bar{n}$ in crystallographic notation. There is no alternative spectroscopic notation because, as shown below, there is an S_n axis corresponding to every $\bar{n}$. The order of the S_n and $\bar{n}$ is not necessarily the same because the S_n axis takes the order corresponding to the rotor-reflection axis $\tilde{n}$.

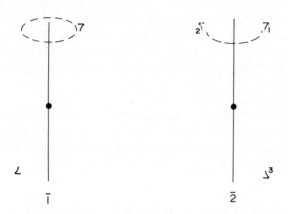

Figure 3.2   1-fold and 2-fold rotor-inversion axes

Figure 3.2 shows the 1-fold and 2-fold rotor-inversion axes. The 1-fold rotor-inversion requires an asymmetric object to be rotated through 360°/1 and then inverted through a point. The 2-fold rotor-inversion axis requires a rotation from position 7_1 through 360°/2 to position 7_2, followed by inversion through a point to position 7_3 (7_1 and 7_3 are therefore related by a 2-fold rotor-inversion axis.) Comparison of Figures 3.1 and 3.2 shows that $\bar{1} \equiv S_2$ ($\tilde{2}$) and $\bar{2}$ is identical with S_1 ($\tilde{1}$). Table 3.1 shows the equivalence of rotor-reflection and rotor-inversion axes, and lists the conventional crystallographic and spectroscopic symbols for these axes. The diagrammatic symbols used in projections are the open polygons with the order of the rotor-reflection axis, as illustrated in Table 3.1.

Table 3.1. Equivalence of rotor-inversion and rotor-reflection axes

Axes		Conventional symbol		
Rotor-reflection	Rotor-inversion	crystallographic	spectroscopic	Diagrammatic
$\tilde{1}$ ≡	$\bar{2}$	m	$S_1 \equiv \sigma$	
$\tilde{2}$ ≡	$\bar{1}$	$\bar{1}$	$S_2 \equiv i$	0
$\tilde{3}$ ≡	$\bar{6}$	$\bar{6}$	S_3	△
$\tilde{4}$ ≡	$\bar{4}$	$\bar{4}$	S_4	□
$\tilde{6}$ ≡	$\bar{3}$	$\bar{3}$	S_6	⬡

Table 3.1 shows, as does comparison of Figures 3.1 and 3.2, that the 1-fold rotor-inversion and the 2-fold rotor-reflection axes are equivalent to a centre of symmetry. In fact $\bar{1}$ is the conventional symbol for a centre of symmetry in the Hermann-Mauguin nomenclature because of its equivalence to a centre of symmetry. Similarly the 2-fold rotor-inversion axis and the 1-fold rotor-reflection axis are equivalent to a plane of symmetry.

The rotor-reflection axis constitutes another symmetry element in terms of which molecular symmetry is to be described. Figure 3.3 shows the configurations generated by the S_3 axis in the trigonal bipyramidal molecule PF_5. We see that, unlike C_3^3, S_3^3 is not equivalent to the identity.

Rule 4(d) on page 21 regarding operations which commute, states that a rotation commutes with a reflection in a plane perpendicular to the rotation axis. Now we can write S_3^1 as $(C_3^1 \times \sigma_h)$ or $(\sigma_h \times C_3^1)$. Therefore:

$$S_3^3 = (C_3^1 \times \sigma_h) \times (C_3^1 \times \sigma_h) \times (C_3^1 \times \sigma_h) = (C_3^3 \times \sigma_h \times \sigma_h \times \sigma_h)$$

Since C_3^3 is equivalent to the identity and $(\sigma_h \times \sigma_h)$ is also equivalent to the identity, we have $S_3^3 \equiv \sigma_h$. By similar reasoning we can see that S_n^n is always equivalent to σ_h when n is odd, and is always equivalent to the identity when n is even. Table 3.2 which is the multiplication table for the operations of the symmetry elements present in allene (see Figure 3.4) illustrates these points.

Figure 3.5 shows the directions of S_n axes ($n = 2, 3, 4$ and 6) in *trans* 1,2-dichloro-1,2-dibromoethane, BF_3, methane and S_6 respectively. Except for *trans* 1,2-dichloro-1,2-dibromoethane, all these molecules possess symmetry elements other than the S_n axis. For example, the S_4 axis in methane is also a C_2, hence the use of the symbol shown in Figure 3.5. Similarly the S_6 in the sulphur allotrope S_6 is also a C_3. The sulphur allotrope S_8 contains an $S_8(C_4)$ axis.

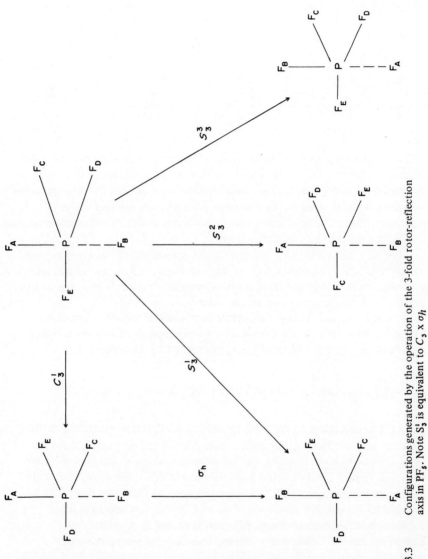

Figure 3.3   Configurations generated by the operation of the 3-fold rotor-reflection axis in PF_5. Note S_3^1 is equivalent to $C_3 \times \sigma_h$

Table. 3.2. Multiplication table for the symmetry operations of allene

	I	$S_4^2 \equiv C_2^1$	S_4^1	S_4^3	$C_{2(1)}$	$C_{2(2)}$	σ_{d1}	σ_{d2}
I	I	C_2^1	S_4^1	S_4^3	$C_{2(1)}$	$C_{2(2)}$	σ_{d1}	σ_{d2}
$S_4^2 \equiv C_2^1$	C_2^1	I	S_4^3	S_4^1	$C_{2(2)}$	$C_{2(1)}$	σ_{d2}	σ_{d1}
S_4^1	S_4^1	S_4^3	C_2^1	I	σ_{d2}	σ_{d1}	$C_{2(1)}$	$C_{2(2)}$
S_4^3	S_4^3	S_4^1	I	C_2^1	σ_{d1}	σ_{d2}	$C_{2(2)}$	$C_{2(1)}$
$C_{2(1)}$	$C_{2(1)}$	$C_{2(2)}$	σ_{d1}	σ_{d2}	I	C_2^1	S_4^1	S_4^3
$C_{2(2)}$	$C_{2(2)}$	$C_{2(1)}$	σ_{d2}	σ_{d_1}	C_2^1	I	S_4^3	S_4^1
σ_{d1}	σ_{d1}	σ_{d2}	$C_{2(2)}$	$C_{2(1)}$	S_4^3	S_4^1	I	C_2^1
σ_{d2}	σ_{d2}	σ_{d1}	$C_{2(1)}$	$C_{2(2)}$	S_4^1	S_4^3	C_2^1	I

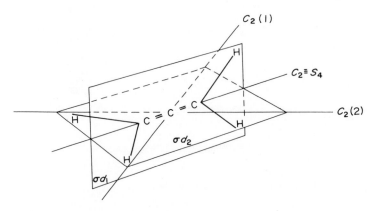

Figure 3.4   Symmetry elements present in the allene molecule

We have seen that molecular symmetry can be defined in terms of the symmetry elements which the molecule possesses. In a molecule, the requirement that the centre of mass remains invariant to any symmetry operation means that all symmetry elements must pass through a point. It is convenient to set up a systematic classification which describes the combination of symmetry elements passing through that point without specifically enumerating them. In order to do this we should first see whether there are any restrictions on the ways in which symmetry elements can be combined in a given system. The allowed combinations of elements constitute the basis of the point group classification.

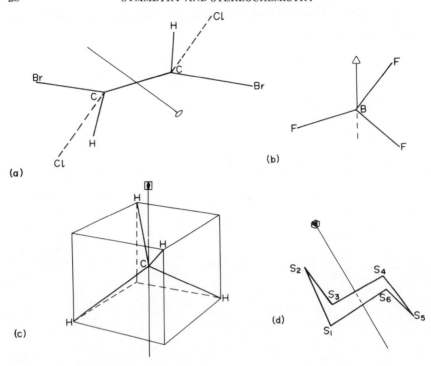

Figure 3.5   Directions of rotor-reflection axes in some molecules
                  (a) S_2 in *trans*-1,2-dichloro-1,2-dibromoethane
                  (b) $S_3 \equiv C_3$ in BF_3
                  (c) S_4 in methane
                  (d) S_6 in the S_6 allotrope of sulphur

Point Group Classification

Rotations, reflections and inversions may be combined in an infinite number of ways, but not all imaginable combinations are possible, nor does every possible combination define a different symmetry. These restrictions arise because of certain interrelations between symmetry elements, as follows:

1. The intersection cf two reflection planes is a symmetry axis. If the angle between the planes is $180°/n$, the axis is of order n.
2. If a reflection plane contains an axis of order n, there must be $n - 1$ other such planes separated by angles of $180°/n$.
3. If there are two C_2 separated by $180°/n$ there must be a C_n perpendicular to them.

4. If there is a C_n with a C_2 perpendicular to it, there must be a further $n - 1$ C_2 axes, separated by $180°/n$.

5. The presence of any two of the elements C_n (n even), σ_h, i automatically implies the third.

In view of the restrictions 1 to 5 above, we cannot have a combination of symmetry elements such as $(C_n + m\sigma_v)$ unless $m = n$. Figure 3.6 shows this for the case $n = 2$.

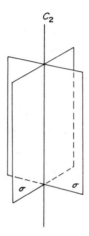

Figure 3.6   Illustration that the intersection of n mirror planes is a C_n axis for the case where $n = 2$

An example of the identity of apparently different descriptions of symmetry involves the symmetry element S_3. By definition, this is a rotation about an axis through $360°/3$ followed by a reflection in a plane perpendicular to the axis; we could, therefore, describe this symmetry by the combination $(C_3 + \sigma_h)$. Thus these two descriptions, S_3 and $(C_3 + \sigma_h)$ define the same symmetry, as Figure 3.3 has already shown. Subject to the above restrictions, a molecule may have any combination of symmetry elements, but the great majority of molecules, and all crystals, have symmetries in which the only allowed values of n, the order of an axis, are 1, 2, 3, 4, 6. If we introduce this third restriction, there are now only 32 possible combinations of symmetry elements, (i.e. of rotation,. reflection, inversion and rotor reflection). These are called the *Crystallographic Point Groups* and they are listed in Table 3.3 along with their Schoenflies and Hermann-Mauguin symbols.

SYMMETRY AND STEREOCHEMISTRY

Table 3.3, The 32 Crystallographic Point Groups

Schoenflies symbol	Hermann-Mauguin symbol	
	Full	Short (minimum elements required to describe group)
C_1	1	1
$C_i(S_2)$	$\bar{1}$	$\bar{1}$
C_s $(C_{1v},\ C_{1h})$	m	m
C_2	2	2
C_{2h}	2/m	2/m
C_{2v}	mm2	mm
$D_2(V)$	222	22
D_2h (V_h)	2/m2/m2/m	mmm
C_4	4	4
S_4	$\bar{4}$	$\bar{4}$
C_4h	4/m	4/m
D_2d (V_d)	$\bar{4}$2m	$\bar{4}$m
C_{4v}	4mm	4m
D_4	422	42
D_4h	4/m2/m2/m	4/mm
C_3	3	3
C_{3i} (S_6)	$\bar{3}$	$\bar{3}$
C_{3v}	3m	3m
D_3	32	32
D_3d	$\bar{3}$2/m	$\bar{3}$m
C_3h	$\bar{6}$	$\bar{6}$
C_6	6	6
C_6h	6/m	6/m
D_3h	$\bar{6}$m2	$\bar{6}$m
C_{6v}	6mm	6m
D_6	622	62
D_6h	6/m2/m2/m	6/mm
T	23	23
T_h	2/m $\bar{3}$	m3
T_d	$\bar{4}$3m	$\bar{4}$3m
O	432	43
O_h	4/m$\bar{3}$2/m	m3m

THE SCHOENFLIES NOTATION

The Schoenflies symbols indicate the elements which generate the group and the 32 crystallographic point groups are made up as follows:

1. Groups whose only symmetry element is a rotation axis (C_1, C_2, C_3, C_4, C_6).
2. Groups whose only symmetry element is a rotor-reflection axis. With the exception of S_4, all of these can be, and usually are, expressed in terms of other combinations of symmetry elements, and will be dealt with under those headings. Groups of the type C_n and S_n are said to be of order n; this means that they possess n symmetry operations (which are generated by the one n-fold axis).
3. Groups consisting of an n-fold rotation axis with n 2-fold axes perpendicular to it. These are the so-called dihedral groups D_n (D_2, D_3, D_4, D_6).
4. Groups consisting of an n-fold rotation axis and n vertical planes of symmetry. These are the C_{nv} groups (C_{2v}, C_{3v}, C_{4v}, C_{6v} and also $C_{1v} \equiv C_s$).
5. Groups consisting of an n-fold rotation axis and a plane perpendicular to it. These are the C_{nh} groups (C_{2h}, C_{4h}, C_{6h} and C_{3h} which is equivalent to S_3).
6. Groups consisting of an n-fold rotation axis and a centre of symmetry. These are the C_{ni} groups (C_i, C_{3i}) which are equivalent to S_{2n} when n is odd.

All groups of the type D_n, C_{nv}, C_{nh} and C_{ni} are of order $2n$, and this is the logic behind regarding S_3 as C_{3h}, since the combination of ($C_3 + \sigma_h$) generates six operations, which is $2n$ as in the other C_{nh} groups.

7. Groups consisting of an n-fold rotation axis, n 2-fold rotation axes and a horizontal reflection plane. These are the D_{nh} groups (D_{2h}, D_{3h}, D_{4h}, D_{6h}).
8. Groups consisting of an n-fold rotation axis, n 2-fold rotation axes and n planes bisecting the angles between the 2-fold axes. These are the D_{nd} groups (D_{2d}, D_{3d}).

D_{nh} and D_{nd} groups are of order $4n$. This arises for D_{nh} because such groups combine the $2n$ operations of the D_n group (n operations generated by the C_n and one by each of the n C_2) with σ_h, which doubles the number of operations again. For D_{nd} the way in which the $4n$ operations are generated from the $2n$ of the D_n group depends on the parity of n. If n is odd, the point of intersection of all the axes is a centre of symmetry and this doubles the number of operations. If n is even, the C_n is also an S_{2n}, producing the $2n$ operations in addition to the $2n$ of the parent D_n group.

We have now accounted for the 27 so-called *Axial* point groups. This name derives from the fact that the symmetry involves one principal axis which is a unique axis, that is, either the only axis or the axis of highest order and the only one of that order.

A difficulty arises here over the D_2, D_{2d} and D_{2h} groups, where there are three mutually perpendicular 2-fold axes. In the D_{2d} group, one can distinguish the unique axis since it is also an S_4 (e.g. the C = C = C direction in allene). The same would apply for the D_2 group, which would be the point group of allene if the angle between the two CH_2 planes were not $0°$ or $90°$. With D_{2h} the unique

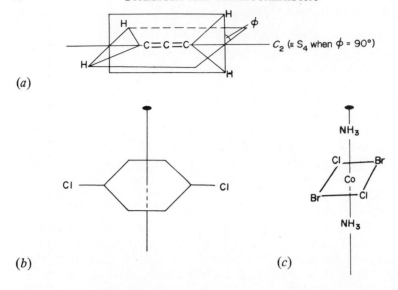

(a)

(b)   (c)

Figure 3.7   The unique 2-fold axis in some molecules with D_2, D_{2d} and D_{2h} symmetry
(a) Unique C_2 axis in allene. When ϕ is 90°, allene has D_{2d} symmetry and
the C_2 is also an S_4. If ϕ were not 0° or 90°, allene would have D_2 symmetry.
(The direction of the unique axis would be the same but it would not be
an S_4 axis)
(b) The axis perpendicular to the plane of the p-dichlorobenzene molecule
is designated as the unique axis. The symmetry of the molecule is D_{2h}
(c) In the hypothetical molecule of D_{2h} symmetry, there are three 2-fold
axes, any one of which could be chosen arbitrarily as the 'unique' axis

axis is taken perpendicular to the molecular plane if the molecule is planar (as
in p-dichlorobenzene), but if the molecule is octahedral $MX_2Y_2Z_2$ the choice
of unique axis becomes arbitrary. Figure 3.7 illustrates these points.

The remaining 5 crystal point groups are the *Cubic* point groups, since they
are based on the rotational symmetry of figures which can be inscribed in a
cube. The simplest of the 5 cubic point groups is the group T, which consists of
the rotation operations of a regular tetrahedron. The symmetry elements are
thus seen to be the four cube diagonals, which are each a C_3, and three 2-fold
axes passing through the centres of opposite faces; these are each a C_2. Since
the four C_3 each generate a C_3^1 and a C_3^2 the group T has 12 symmetry
operations. The 3-fold axes are shown in Figure 3.8.

A tetrahedral molecule XY_4 belongs to the group T_d; as well as the twelve
rotation operations there are now 6 planes of symmetry, each of which contains
two atoms and bisects the angle between the bonds to the other two. The three
C_2 now also become S_4 which each generate the operations S_4^1 and S_4^3 ($S_4^2 \equiv C_2$

Figure 3.8   A tetrahedron inscribed in a cube. The four cube diagonals are the four
                  3-fold axes which represent the minimum symmetry for any cubic figure.
                  If a tetrahedral molecule XY_4 were placed in the cube with X at point A a
                  possible set of positions for the Y atoms is on corners 1, 2, 3, 4

so does not count as a separate operation). There are thus 24 operations in the
group T_d.

The other group based on the rotation operations of a tetrahedron is T_h,
which is obtained by combining the 12 rotation operations of T with a centre
of symmetry, giving a total of 24. The group O is also of order 24, consisting of
the rotation operations of a regular octadehron. The axes which were 2-fold axes
in the group T now become 4-fold axes in addition, as Figure 3.9 shows.

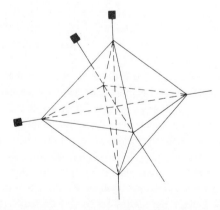

Figure 3.9   Rotation axes of an octahedron. The three C_4 axes through opposite vertices
                  of the octahedron and the mid positions of the cube faces are also C_2; the
                  other C_2 axes pass through the mid points of opposite edges of both the
                  octahedron and the cube in which it is inscribed

There is also another set of 2-fold axes, which pass through the centres of opposite edges of the octahedron and of the cube in which it is inscribed. Finally, if the elements of the group O are combined with a centre of symmetry, the group O_h, which is the point group of all octahedral molecules XY_6, such as SF_6, is obtained. The 48 operations of the group now include 9 planes of symmetry and the three C_4 also become S_4, while the four C_3 become S_6.

The alternative designations among the Schoenflies symbols in Table 3.3 arise for a variety of reasons:

1. D_2 is sometimes designated V (German, *Vierergruppe*).
2. S_{2n} is equivalent to C_{ni} when n is odd.
3. In the group whose only symmetry elements are the identity and a plane of symmetry (which is necessarily the molecular plane when the molecule is planar), the C_1 axis which represents the identity can be considered as either perpendicular to, or contained in, that plane. The plane is thus either a σ_h or a σ_v, so the group can be called C_{1h} or C_{1v}, but is usually called C_s.

THE HERMANN-MAUGUIN NOTATION

The Hermann-Mauguin symbols for any point group is a list of the symmetry elements which describe the group. The short form lists the minimum number of symmetry elements required to completely define that point group. If we follow the same pattern that we used to describe the Schoenflies notation we have the following Hermann-Mauguin groups in their full form.

1. Groups whose only symmetry element is a rotation axis (1, 2, 3, 4, 6).
2. Groups whose only symmetry element is a rotor-reflection axis ($\bar{1}$, $\bar{2} \equiv$ m, $\bar{3}$, $\bar{4}$, $\bar{6}$). The symbol m for mirror plane is used in preference to $\bar{2}$.
3. Groups consisting of an n-fold rotation axis within n 2-fold axes perpendicular to it (222, 32, 422, 622). Note it is not necessary to specify all of the 2-fold axes because operation of an n-fold axis on one 2-fold axis perpendicular to it generates all n 2-fold axes. Thus the shortened forms of these symbols for these groups are designated $n2$.
4. Groups consisting of an n-fold rotation axis and n vertical planes of symmetry (mm2, 3m, 4mm, 6mm); again the short form symbols are the minimum number of elements required to describe the group. For example a 2-fold axis and two vertical mirror planes should logically be 2mm but as we have seen in Figure 3.6 the presence of any two of these elements (i.e. mm or 2m) implies the third. mm is conventionally taken as the short symbol for this group and the 2 is taken as the redundant element as far as describing the group is concerned.
5. Groups consisting of an n-fold rotation axis and a plane perpendicular to it. The symbol used for a plane perpendicular to an n-fold axis is n/m and the groups are $2/m$, $3/m \equiv \bar{6}$, $4/m$, $6/m$. The symbol $\bar{6}$ is preferred to $3/m$ as the description of the group.
6. Groups with n-fold rotation axes and a centre of symmetry. The groups in this

Schoenflies list, C_i and C_{3i}, are the groups $\bar{1}$ and $\bar{3}$ respectively in Hermann-Mauguin notation.

7. Groups consisting of an n-fold rotation axis, n 2-fold axes and a horizontal reflection plane (mmm, 3/mm $\equiv \bar{6}$m2, 4/mmm, 6/mmm) $\bar{6}$m2 is preferred to 3/mm as the symbol for this group. Again the short forms listed are the necessary elements.

8. Groups consisting of an n-fold rotation axis, n 2-fold rotation axes and n planes bisecting the angles between the 2-fold axes. These are the Schoenflies groups D_{2d} and D_{3d} for which the Hermann-Mauguin symbols are $\bar{4}$m and $\bar{3}$m.

9. The Hermann-Mauguin symbols for the cubic point groups are 23, 2/m $\bar{3}$ ($\equiv$ m3), 43m, 432 and m/4 $\bar{3}$ m/2 ($\equiv$ m3m) which had the Schoenflies symbols T, T_h, T_d, O and O_h respectively.

The Hermann-Mauguin symbols for the 27 axial point groups fall into seven classes with general symbols n, $\bar{n}$, $n2$, nm, n/m, n/mm, $\bar{n}$m. If we take n as being 1, 2, 3, 4 or 6 it would appear that there are 35 of these point groups. This number, however, reduces to 27 as can be seen from Table 3.4 because certain of the possible combinations are equivalent to other point groups. This provides another illustration of the statement made earlier in this chapter that not all possible combinations of symmetry elements define different symmetries.

Table 3.4. The possible axial point groups in the Hermann-Mauguin nomenclature

General point-group class	Possible point-group with n-fold axis, (n = 1, 2, 3, 4, 6).				
	1	2	3	4	6
n	1	2	3	4	6
$\bar{n}$	$\bar{1}$	$(\bar{2} \equiv m)$	$\bar{3}$	$\bar{4}$	$\bar{6}$
$n2$	$(12 \equiv 2)$	22	32	42	62
nm	$1m = m$	$2m = mm$	3m	4m	6m
n/m	$(1/m \equiv 2)$	$2/m$	$(3/m \equiv \bar{6})$	$4/m$	$6/m$
n/mm	$(1/mm \equiv 2m)$	$2/mm = mmm$	$(3/mm \equiv \bar{6}m)$	$4/mm$	$6/mm$
$\bar{n}$m	$(\bar{1}m \equiv 2/m)$	$(\bar{2}m \equiv 2m)$	$\bar{3}$m	$\bar{4}$m	$\bar{6}$m

Allocation of Molecules to their Point Groups

We have so far described the 32 crystallographic point groups in which the necessity for translation symmetry in a crystal restricts the possible values of n to 1, 2, 3, 4 and 6 for both rotation and rotor-reflection axes. For molecules this restriction does not apply and, accordingly, point groups involving axes of any order are allowed. No new general type of point group arises, so that we still have to assign any given molecule to a point group of the type C_n, S_n, D_n, C_{nv}, C_{nh}, D_{nd}, D_{nh} (axial point groups) or to one of the five cubic point groups T, T_h, T_d, O, O_h. Most molecules belong to one of the 32 crystallographic point groups, and Tables 3.5 and 3.6 list these, with examples of real or hypothetical molecules belonging to them. Some molecules belong to one of

SYMMETRY AND STEREOCHEMISTRY

Table 3.5. Molecules and ions belonging to one of the 27 axial crystallographic point groups

Type	n	Shape	Molecule or ion	Remarks
C_n	1		CHFClBr	
	2		H_2O_2	
	3		1, 1, 1-trichloroethane	Provided the two ends are gauche
	4		not known	Molecules $X(YZ)_n$ would
	6		not known	have such point groups if they were non-planar, with all $X\hat{Y}Z$ equal, but not $180°$
S_n	4		3, 4, 7, 8-tetra-methyl-1-aza-spiro-[4, 4]-nonane	(see Figure 3.12)
D_n	2		⎫	Molecules X_2Y_{2n} would
	3		⎬ not known	have such point groups if the two Y_n groups were
	4		⎭	gauche. e.g. ethane in the gauche form would have the point group D_3
	6			
C_{nv}	1	planar but non-linear	HOCl	$C_{1v} = C_s$
		pyramidal	$SOCl_2$	
	2	planar V shape	H_2O	
		planar Y shape	H_2CO	
		planar T shape	ClF_3	
		derived from plane square	cis-$Pt(NH_3)_2Cl_2$	
		derived from tetrahedron	CH_2Cl_2	
		derived from trigonal bipyramid	SF_4	One equatorial position is unoccupied
		derived from hexagon	chlorobenzene	
	3	pyramidal	NH_3	
	4	pyramidal	$XeOF_4$	The Xe–O bond lies along the C_4 axis
		derived from octahedron	SF_5Cl	The S–Cl bond lies along the C_4 axis
	6	not known		
C_{nh}	2		$trans$-N_2F_2	
	3(= S_3)		H_3BO_3	This molecule is planar with all the $B\hat{O}H$ angles equal but not $180°$
	4		not known	
	6		not known	

C_{ni}	1 ($C_i = S_2$)		trans- HClBrC—CBrHCl	
	3		$[Co(NO_2)_6]^3$	This is the symmetry of the ion in the Na salt. Opposite NO_2 groups are gauche.
D_{nh}	2	derived from plane square	trans- $Pt(NH_3)_2 Cl_2$	
		derived from octahedron	trans-$[Co(NH_3)_2 Cl_2 Br_2]$	
		derived from hexagon	p-dichlorobenzene	
		$X_2 Y_{2n}$ molecule with the Y_n groups eclipsed	$B_2 Cl_4$ (solid)	
	3	equilateral triangle	BCl_3 (vapour); NO_3^-	
		derived from hexagon	1,3,5-trichlorobenzene	
		trigonal bipyramid	PCl_5 (vapour)	
	4	plane square	XeF_4	
		derived from octahedron	trans- $[Co(NH_3)_4 Cl_2]^+$	
	6	plane hexagon	Benzene	
D_{nd}	2	$\left.\begin{array}{l} X_2 Y_{2n} \text{ molecule} \\ \text{with } Y_n \text{ groups} \\ \text{staggered} \end{array}\right\}$	$B_2 Cl_4$ (vapour)	
	3		$C_2 H_6$	

Table 3.6. Molecules and ions belonging to one of the 5 cubic point groups

Group	Molecule or ion	Remarks
T	Not known	The CH_3 groups of $C(CH_3)_4$ can be rotated so as to reduce the symmetry of the molecule from T_d to T
T_h	$[(Co(NO_2)_6]^{3-}$	This is the ideal symmetry of the ion and occurs when opposite pairs of nitro groups are eclipsed
T_d	CCl_4 ; $C(CH_3)_4$	
O	Not known	
O_h	SF_6	

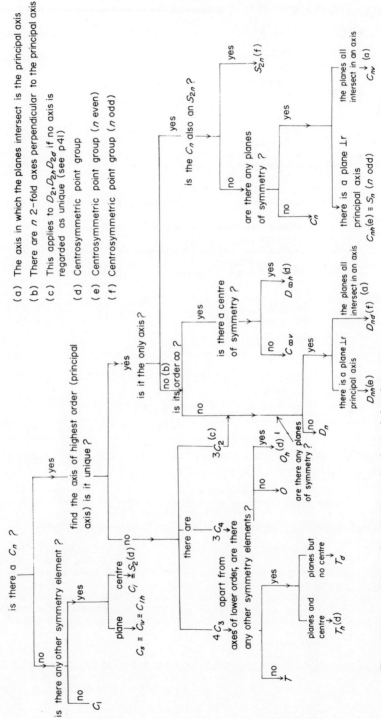

(a) The axis in which the planes intersect is the principal axis

(b) There are n 2-fold axes perpendicular to the principal axis

(c) This applies to D_2, D_{2h}, D_{2d} if no axis is regarded as unique (see p41)

(d) Centrosymmetric point group

(e) Centrosymmetric point group (n even)

(f) Centrosymmetric point group (n odd)

Figure 3.10   Scheme for assigning molecules to their point groups

Table 3.7. Molecules and ions belonging to a non-crystallographic point group

Type	n	Shape	Molecule or ion	Remarks
C_{nv}	5	Pentagonal bipyramid XY_7 with the central X atom out of the equatorial plane		If the I atom of IF_7 were moved out of the equatorial plane the symmetry of the molecule would be reduced from D_{5h} to C_{5v}
	∞	Linear	N_2O	Any linear molecule without a centre of symmetry has this group
D_{nh}	5	Pentagonal bipyramid XY_7 with X in the equatorial plane	ReF_7, IF_7	
		Plane pentagon	Cyclopentadienyl anion	
		Pentagonal prism	Ruthenocene	
	7	Plane heptagon	Cycloheptatrienyl (tropylium) cation	
	∞	Linear	CO_2; C_2H_2; C_3O_2 if linear	Any linear molecule with a centre of symmetry has this group.
D_{nd}	4	Puckered octagon	Sulphur (S_8), cyclo-octatetraene ('crown' form)	
	5	Pentagonal antiprism	Ferrocene	

the non-crystallographic point groups; a list of the more important of these groups is given in Table 3.7.

Figure 3.10 gives a scheme for assigning molecules to their point groups, and we now consider some examples of these assignments, which are made by answering the questions in this figure. Figure 3.10 uses only the Schoenflies notation but the Hermann-Mauguin symbols are also given for the specific examples considered and in some of the other tables.

1. *A molecule whose point group is of the type $C_n(n)$; H_2O_2*

The elements present in H_2O_2 are shown in Figure 3.11.

(a) Is there a C_n?   YES

The C_n of highest order is the C_2 through the mid-point of the O–O bond and bisecting the angle between the two HOO planes.

(b) Is it unique?   YES   (there is no other 2-fold axis)
(c) Is it the only axis?   YES
(d) Is the C_2 also an S_{2n}?   NO
(e) Are there any planes of symmetry?   NO

This sequence of answers shows that the molecule belongs to a C_n point group and, since n is 2, *the point group of H_2O_2 is $C_2(2)$*.

2 = C_2 (passes through mid point of
O-O bond and bisects ϕ)
H
O
θ
O
θ
ϕ
H

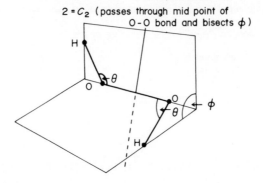

Figure 3.11 Symmetry elements of H_2O_2

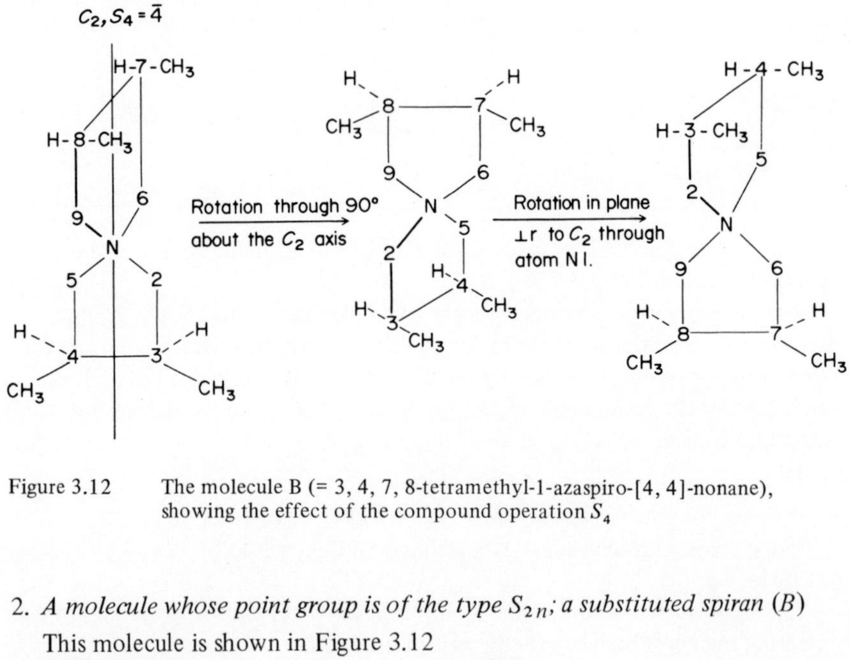

$C_2, S_4 = \bar{4}$
Rotation through 90° about the C_2 axis
Rotation in plane ⊥r to C_2 through atom N I.

Figure 3.12 The molecule B (= 3, 4, 7, 8-tetramethyl-1-azaspiro-[4, 4]-nonane), showing the effect of the compound operation S_4

2. *A molecule whose point group is of the type S_{2n}; a substituted spiran* (*B*)

This molecule is shown in Figure 3.12

(a) Is there a C_n? YES

The C_n of highest order is the C_2 bisecting the two rings which lie in planes at right-angles.

(b) Is it unique? YES

(c) Is it the only axis? YES

(d) Is the C_n also an S_{2n}? YES

It is an S_4. Since the two rings lie in planes at right-angles, rotation through 90° interchanges the orientation of the top and bottom rings; then the reflection at the plane through the central atom, perpendicular to the axis, exchanges the rings again.

Since the only symmetry element is a C_2 which is also an S_4, the molecule B *belongs to the point group* S_4 $(\bar{4})$.

3. *Molecules whose point groups are of the type* D_n

(i) Allene in the 'gauche' form. (This is a hypothetical molecule.)

If the two planes containing the methylene groups in allene are rotated with respect to one another by some angle other than 0° (eclipsed form) or 90° (staggered form) then we have the so-called 'gauche' form. Figure 3.13 shows this form in two views, illustrating the two different types of C_2 axis.

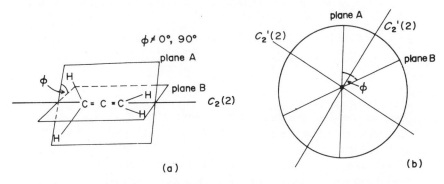

(a)	(b)

Figure 3.13   (a) Perspective view of 'gauche' allene, with the C_2 axis along the C=C=C direction
(b) End view of 'gauche' allene, showing the directions of the C_2' axes

(a) Is there a C_n?   YES
(b) Is it unique?   ?

This point demands some discussion. Although there are only three axes, all of the same order (2), there is a sense in which we can say that the C_2 axis along the C=C=C direction is unique, because the atoms which are invariant to rotation about this axis (the three carbon atoms) are not the same as those invariant to rotation about the C_2' axes (only the central carbon). Distinctions of this nature can be made for the 3 2-fold axes in molecules of the point groups D_{2d} and D_2, but not always for those of the point group D_{2h}. Figure 3.14 illustrates this. We have decided to neglect these finer distinctions, and use 'unique' in the stricter sense of being the only axis of that order, since this enables us to treat all molecules of the point group D_{2h} on the same footing: thus the answer to question (b) is NO.

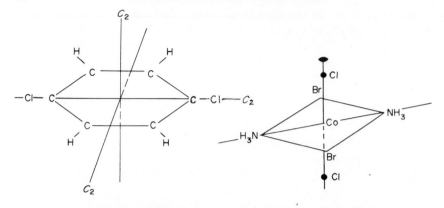

Figure 3.14   (a) In p-dichlorobenzene the C_2 axis through the two Cl atoms leaves 4
atoms invariant to rotation around it, and could therefore be designated as
the unique axis, since the other two C_2 axes leave no atoms invariant to
rotation
(b) In this complex, each of the three 2-fold axes has 3 atoms invariant to
rotation about it, so there is no sense in which any of the axes is unique

There are, in fact, three C_2 axes.

(c) Are there any planes of symmetry?   NO

The molecule thus possesses only three C_2 axes and *belongs to the point
group D_2(222)*.

(ii) Ethane in the 'gauche' form. (This is a hypothetical molecule.)

Here the ambiguity about the unique axis does not arise.

(a) Is there a C_n?   YES

The axis along the C–C direction is a C_3(3).

(b) Is it unique?   YES

(c) Is it the only axis?   NO

There are three other axes, each passing through the mid-point of the C–C
bond and lying in planes which bisect the angles ϕ_1, ϕ_2, ϕ_3, shown in Figure 3.15.

(d) Is its order ∞?   NO

(e) Are there any planes of symmetry?   NO

The 'gauche' form of ethane *belongs to the point group D_3 (32)*.

4. *A molecule whose point group is of the type D_{nd}*

The reader may verify, by repeating the steps of example 3, that the normal
forms of allene and ethane belong to the point groups D_{2d} and D_{3d} respectively.
A molecule of a rather different geometrical type with a D_{nd} type point group
is the puckered hexagonal molecule formed by the S_6 allotrope of sulphur.

(a) Is there a C_n?   YES

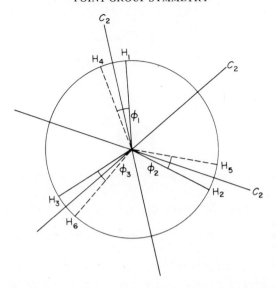

Figure 3.15   End-on view of the 'gauche' form of the ethane molecule. ϕ_1 defines the angle between the planes containing H_1 and the two carbons, and the two carbons and H_4. A C_2 lies in the plane bisecting ϕ_1, and passes through the midpoint of the C–C bond

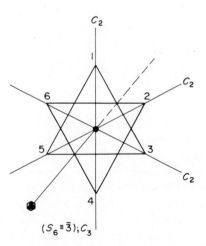

Figure 3.16   Plan view of S_6 molecule, with two planes defined by atoms 1, 3, 5 and 2, 4, 6 respectively. C_2 axes pass through atoms 1 and 4, 2 and 5, 3 and 6; the planes containing these axes, and perpendicular to the planes of atoms 1, 3, 5 and 2, 4, 6 are planes of symmetry

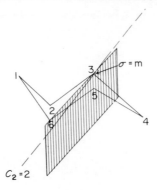

Figure 3.17   A plane of symmetry and a 2-fold axis in the S_6 molecule

It is the 3-fold axis passing through the ring in a direction perpendicular to the planes defined by atoms 1, 3, 5 and 2, 4, 6 as shown in Figure 3.16.

(b) Is it unique?   YES
(c) Is it the only axis?   NO
(d) Is its order ∞?   NO
(e) Are there any planes of symmetry?   YES (see Figure 3.17)

These planes all intersect in the C_3 axis, so *the S_6 molecule belongs to the point group D_{3d}.* ($\bar{3}2/m$).

5. *A molecule whose point group is of the type D_{nh}; 1, 3, 5-trichlorobenzene*

(a) Is there a C_n?   YES

There is a 3-fold axis passing through the centre of the benzene ring, perpendicular to the plane of the molecule.

(b) Is it unique?   YES
(c) Is it the only axis?   NO

There are three C_2 lying in the plane of the ring, each passing through one chlorine and one hydrogen atom.

(d) Are there any planes of symmetry?   YES

There are four, of which three lie in the direction of the C_2, and are perpendicular to the plane of the ring. They intersect in the C_3 axis and are therefore three σ_v. The fourth is the molecular plane, which is perpendicular to the C_3 axis, and is therefore a σ_h. 1, 3, 5-trichlorobenzene *belongs to the point group D_{3h}* ($\bar{6}m2$).

6. *A molecule whose point group is of the type C_{nh}; H_3BO_3*

Figure 3.18 shows the H_3BO_3 molecule. It is planar, with all the $B\hat{O}H$ angles equal but not $180°$. All of the $O\hat{B}O$ angles are equal.

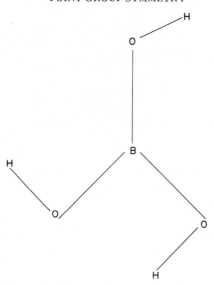

Figure 3.18   The H_3BO_3 molecule

(a) Is there a C_n?   YES
    As all the OB̂O angles are equal, and all the BÔH angles are equal, there must
be a C_3, through the B atom, perpendicular to the molecular plane.
(b) Is it unique?   YES
(c) Is it the only axis?   YES
(d) Is the C_n also an S_{2n}?   NO
    It is, in fact, an S_3 ($\bar{6}$).
(e) Are there any planes of symmetry?   YES
    There is one, the molecular plane. It is important to realise that the plane of a
planar molecule is a plane of symmetry — this seems so trivial that it is often
forgotten when attempting to assign molecules to their point groups. Since this
plane is perpendicular to the C_n axis, it is a σ_h and H_3BO_3 *belongs to the point
group* C_{3h}. ($\bar{6}$).

7. *Molecules belonging to point groups of the type* C_{nv}; CH_2Cl_2 *and* SF_4
    Some types of molecule are easily recognisable as having C_{nv} symmetry.
Pyramidal molecules XY_n where the lines joining the Y atoms enclose a
regular n-sided figure constitute one such type; examples are NH_3, PCl_3. If an
atom Z is added to such an arrangement, the X—Z bond being in the direction
of the C_n axis, the symmetry is again C_{nv}. Such molecules include CH_3Cl and

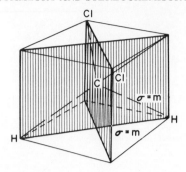

Figure 3.19   The symmetry elements of CH_2Cl_2

$XeOF_4$. The C_{nv} symmetry of the molecules CH_2Cl_2 and SF_4 is perhaps less intuitively obvious.

(i) CH_2Cl_2

(a) Is there a C_n?   YES

It is a C_2 and bisects both the $H\hat{C}H$ and $Cl\hat{C}Cl$ angles (see Figure 3.19).

(b) Is it unique?   YES

(c) Is it the only axis?   YES

(d) Is the C_n also an S_n?   NO

(e) Are there any planes of symmetry?   YES

The plane containing the CCl_2 group and bisecting the $H\hat{C}H$ angle, and that containing the CH_2 group and bisecting the $Cl\hat{C}Cl$ angle, are both planes of symmetry which contain the C_2 axis, and are therefore two σ_v. CH_2Cl_2 *therefore belongs to the point group C_{2v} (mm2).*

(ii) SF_4

SF_4 is now considered to have a trigonal bipyramidal shape, with one of the equatorial positions unoccupied (see Figure 3.20).

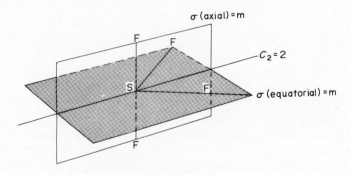

Figure 3.20   The symmetry elements of SF_4

(a) Is there a C_n?   YES

It is a C_2 and bisects the $F\hat{S}F$ (equatorial) and the $F\hat{S}F$ (axial) angles.

(b) Is it unique?   YES

(c) Is it the only axis?   YES

(d) Is the C_2 also an S_{2n}?   NO

(e) Are there any planes of symmetry?   YES

As for CH_2Cl_2, there are $2\sigma_v$. In SF_4, one σ_v contains the equatorial, and one the axial fluorine atoms. The C_2 is their line of intersection. SF_4 *belongs to the point group* C_{2v} (mm2).

8. *Molecules belonging to the cubic point groups;* CCl_4 and SF_6

The above two molecules represent the most important of the cubic point groups, as far as molecular symmetry is concerned, though a few molecules with T_h symmetry are known.

In CCl_4, there are four 3-fold axes, directed along the C–Cl bonds; these constitute the minimum requirements for cubic symmetry. If the molecule were inscribed in a cube with the Cl atoms at the alternate corners, the four $C_3(3)$ axes would lie in the direction of the cube diagonals. In addition to the four C_3, there are three C_2 (2) which are also S_4 ($\bar{4}$). In terms of the molecular symmetry, they would bisect the opposite pairs of angles; in terms of the symmetry of a cubic figure, they join the centres of opposite faces. In a cube with all eight corners identical, they would, of course, be 4-fold axes. If we look at the CCl_4 molecule, we find symmetry elements other than the rotation axes; each of the six planes containing two chlorines and a carbon is a plane of symmetry. ($\sigma_d \equiv m$) There is, however, no centre of symmetry, so CCl_4 *belongs to the group T_d*. (43m).

In SF_6 the lines joining opposite fluorine atoms (of which there are three pairs) are the three $C_4(4)$ axes; these are also $C_2(2)$ axes, since a configuration equivalent to the original is obtained by a rotation through either 90° or 180° about these axes. Further, these axes are also S_4 axes, as Figure 3.21 shows. The S atom is situated at a centre of symmetry and there are nine planes of symmetry (m). Three of these are σ_h, containing four fluorine atoms as well as the sulphur; the other six, which are σ_d, bisect FSF angles and contain only the sulphur atoms. Since SF_6 contains planes of symmetry and a centre of symmetry as well as the three C_4 axes, it *belongs to the point group O_h*, (4/m $\bar{3}$ 2/m).

The position of the $S_6 \equiv \bar{3}$ axis used in the full Hermann-Mauguin notation is shown in Figure 3.22.

Two points, which have only been implied so far, should now be made explicitly.

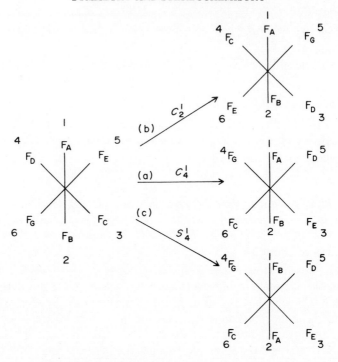

Figure 3.21   Different configurations produced when the axis $1-2$ in SF_6 is treated as (a) a C_4 (b) a C_2 (c) an S_4

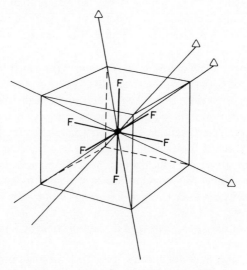

Figure 3.22   The positions of the $S_6 \equiv \bar{3}$ axes in octahedral SF_6

1. It is not necessary, in general, to identify every symmetry element in the molecule to assign it to its correct point group. If, for instance, the reader goes through the argument of example 3, he will see that it is not necessary to identify the n C_2 axes in order to assign the hypothetical gauche forms of allene or ethane to the groups D_2 and D_3 respectively. It will, however, in later applications, be necessary to identify these elements.

2. In many cases, the same axis can correspond to more than one symmetry element. Sometimes, as in example 2, where we found the same axis to be a C_2 and an S_4, it is necessary to realise that such a situation exists in order to assign the correct point group. In other cases, such as in SF_6, where the C_4 axes are also C_2 and S_4, and in tetrahedral CCl_4, where the C_2 axes are also S_4, it is not necessary. Again, there will be applications considered later where it is important to realise that the same axis corresponds to more than one symmetry element.

Reduction of Symmetry

As we shall see in Chapter 6, many of the physical properties of a molecule depend on its symmetry. The chemist makes use of this dependence in two ways. Firstly, if he is studying a molecule about whose shape he knows nothing, he can work out all its possible configurations, assign a point group to each of them, and then make predictions about the physical properties to be associated with each suggested configuration. Secondly, he may be studying a molecule of known symmetry and may wish to determine the nature of the products found in various reactions involving it. In many cases, particularly involving co-ordination compounds, possible products may have the same composition but different symmetry. For instance we may convert the hexamminocobalt(III) ion to either the *cis-* or *trans-* dichlorotetramminocobalt(III) ion, and the best way to determine which of these has been formed is to study those physical properties which depend on the symmetry of the two possible products. In the light of this application, it is clearly useful to be able to work out how the symmetry of a molecule is affected by making originally equivalent features non-equivalent. We may recognize, in principle, two ways of doing this:

1. A distortion of the molecule involving changes in bond lengths and/or bond angles.

2. Replacement of atoms or groups of atoms by others which are not identical to them.

In general, the second of these effects will imply the first, but the converse is not necessarily true, since a distortion of the molecule may arise simply from the application of a field. In order to illustrate the idea of reduction in symmetry, we shall consider both types of effect.

We can consider three high-symmetry point groups from which all the others can be obtained by various distortions. These three groups are:

(i) the octahedral group O_h

(ii) the tetrahedral group T_d

(iii) the group of the plane hexagon, D_{6h}.

Although all the symmetry elements of T_d are also elements of O_h (we say that T_d is a *Sub-Group* of O_h, a term which will be explained in Chapter 5) it is not really convenient, particularly in terms of molecular symmetry, to think of T_d as being obtainable from O_h by distortion. Figure 3.23 illustrates this by showing cubes with atoms at the corners of the polyhedra.

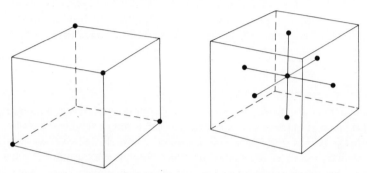

Figure 3.23   A cube with atoms disposed so as to give octahedral or tetrahedral symmetry

In considering the ways in which the symmetry of a given molecule or array of atoms is reduced by distortion, we need to determine which symmetry elements are destroyed by, and which are retained in spite of, the distortion. We shall illustrate this by means of simple geometrical figures as well as molecular shapes.

REDUCTION OF SYMMETRY FROM O_h

1. *By tetragonal distortion*

A distortion along one of the 4-fold axes, as shown in Figure 3.24, will leave that axis as the only C_4; the symmetry will be reduced to that of a plane square (D_{4h}), all the 3-fold axes and the elements associated with them being destroyed.

The symmetry of the plane square can be then reduced in a number of ways, shown in Figure 3.25.

(a) Reductions in which the centre of symmetry is retained.

If the figure formed by the two diagonals is considered, the symmetry can be reduced by making the diagonals unequal in length (D_{2h}), or by adding feet to each of the diagonals to give a swastika shape (C_{4h}).

(b) Reductions in which the centre of symmetry is not retained, but some planes of symmetry are kept.

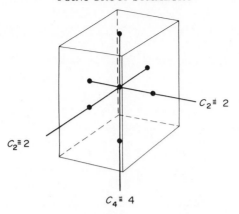

Figure 3.24   Effect of tetragonal distortion of a cube giving a figure with D_{4h} symmetry

If the four half-diagonals are kept equal in length, but their point of inter-section is moved out of the plane, the symmetry becomes that of a square-based pyramid (C_{4v}).

If a plane square is folded along one diagonal the resultant figure is then a puckered four-membered ring and each diagonal lies in a plane of symmetry. The axis which was both a C_4 and an S_4 in the plane square is now a C_4 only, and the symmetry of the figure is D_{2d}.

(c) Reductions in which only symmetry axes are retained.

If two plane squares are superimposed so that their centres are coincident and the diagonals of one do not coincide with those of the other, nor bisect the angles formed by them, the symmetry elements retained are one C_4 and four C_2, giving the point group D_4 .

2. By rhombohedral distortion

If an octahedron is elongated in the direction of one of the C_3 axes, this becomes the only such axis and all the symmetry elements involving the 4-fold axes are destroyed. The point group resulting from such a distortion is D_{3d}. The solid figure with this symmetry is the trigonal antiprism and its symmetry can be reduced in two ways, neither of which retains the centre of symmetry.

   (i) If the face ABC of the trigonal antiprism is rotated through some angle (other than 60° or an integral multiple of it) relative to the face DEF, the C_3 is no longer an S_6 and the three σ_d disappear, leaving a figure of D_3 symmetry.

   (ii) If we replace the face ABC by a point G at its centre, and join D, E and F to this point, we have a triangular-based pyramid of C_{3v} symmetry. The C_3 is no longer an S_6 and the 2-fold axes are destroyed.

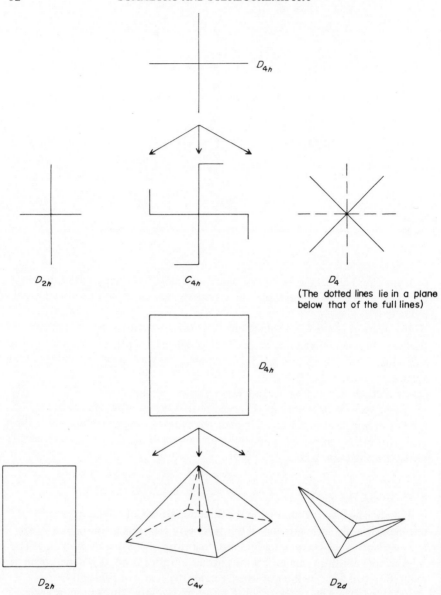

D_{4h}

D_{2h} C_{4h} D_4
(The dotted lines lie in a plane
below that of the full lines)

D_{4h}

D_{2h} C_{4v} D_{2d}

Figure 3.25   Figures with D_{4h} symmetry and their reduction to lower symmetries

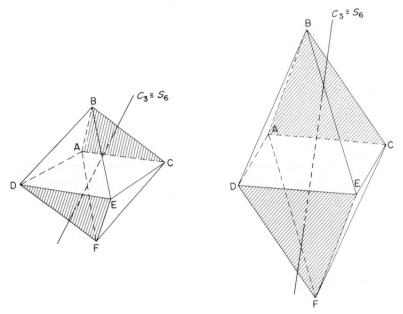

Figure 3.26   A regular octahedron (left) and the trigonal antiprism of D_{3d} symmetry (right), produced by elongating the octahedron along one of the C_3 axes

Another figure with D_{3d} symmetry is the puckered hexagon shown in Figure 3.16. The symmetry of this can be reduced to S_6 by adding bonds orientated in such a way that the three 2-fold axes and three planes of symmetry are destroyed, but the centre of symmetry and the S_6 axis are retained. Figure 3.27 shows this reduced symmetry.

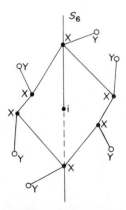

Figure 3.27   Puckered hexagon X_6Y_6 with S_6 symmetry. If the six X-Y bonds are removed, the figure has D_{3d} symmetry

REDUCTION OF SYMMETRY FROM T_d

1. *By tetragonal distortion*

Consider the tetrahedron inscribed in a cube, as shown in Figure 3.8. The axes joining opposite face-centres are S_4 and the cube diagonals are C_3. If we elongate the cube in the direction of one of the S_4 axes, this becomes the only S_4 axis and the 3-fold axes are destroyed. The symmetry of the resulting figure is D_{2d}. One can see this by comparing the shapes of methane and allene, which are shown in Figure 3.28, as inscribed in a cube and in a tetragonal prism respectively.

A further reduction of the D_{2d} symmetry can be effected by rotating one of the CH_2 groups in allene relative to the other; this leads to D_2 symmetry since the axis lying the in C–C–C bond direction is no longer an S_4 axis and the planes containing the CH_2 groups are no longer σ_d.

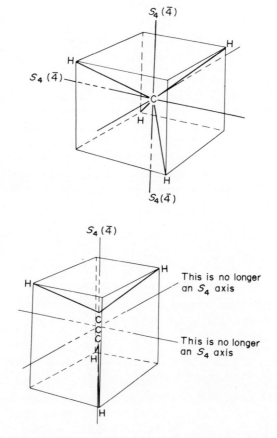

Figure 3.28   Methane inscribed in a cube, and allene inscribed in a tetragonal prism

2. *By pyramidal distortion*

If the cube in which the tetrahedron is inscribed is elongated along one of the cube diagonals all the 4-fold axes are destroyed and the direction of elongation becomes the only C_3 axis. The only planes of symmetry retained each contain the C_3 axis and one of the other bonds, so that the distortion results in C_{3v} symmetry.

If the cube is distorted in the directions of the diagonals such that there are two sets of two equivalent directions, the symmetry is reduced to C_{2v}, and if the distortion is such as to lead to one set of two equivalent directions and to two other directions which are not equivalent to those or to each other, the symmetry is reduced to C_s. Figure 3.29 illustrates these reductions in symmetry.

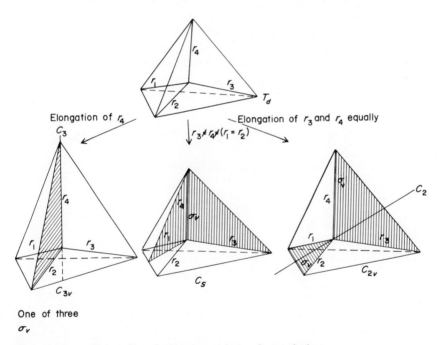

Figure 3.29   Progressive reduction in symmetry of a tetrahedron

REDUCTION OF SYMMETRY FROM D_{6h}

We may consider the possible distortions of the plane hexagon, which retain a 3-fold or 6-fold axis, in terms of the retention or destruction of two symmetry elements, namely, the plane of the hexagon, which is a σ_h, and the centre of symmetry i. Table 3.8 shows the results of the possible distortions.

Table 3.8. Results of reducing the symmetry of a plane hexagon

Element	Retained		Destroyed		Resultant point group
	i	σ_h	i	σ_h	
	✓	✓			C_{6h}
	✓			✓	D_{3d}
		✓	✓		D_{3h}
			✓	✓	C_{6v} or D_6

C_{6h} is a point group not normally found; Figure 3.30 shows how it may be derived from a figure of D_{6h} symmetry. The common point group D_{3h}, as well as being derived from the plane hexagon, occurs in the form of two three-dimensional figures, the trigonal prism and the trigonal bipyramid, shown in Figure 3.31. The first of these is more often found as a co-ordination polyhedron (for example, the AlO_6 arrangement in some alums) than as a discrete molecule, but the trigonal bipyramid and some of the lower-symmetry figures derived from it are common molecular shapes, as Table 3.5 and Figures 3.20 and 3.32 show.

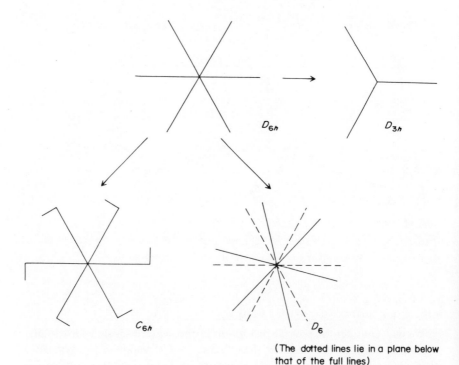

(The dotted lines lie in a plane below that of the full lines)

Figure 3.30   Some possible reductions in symmetry of a figure whose point group is D_{6h}

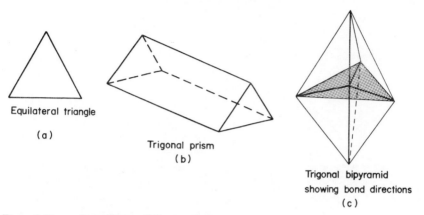

Equilateral triangle

(a)

Trigonal prism
(b)

Trigonal bipyramid
showing bond directions
(c)

Figure 3.31   Some figures of D_{3h} symmetry
                 (a) equilateral triangle
                 (b) trigonal prism
                 (c) trigonal bipyramid showing bond directions

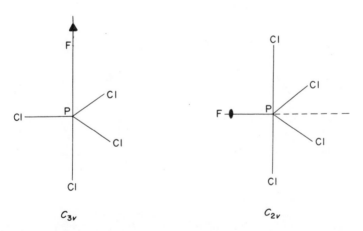

C_{3v} C_{2v}

Figure 3.32   Possible configurations of PCl_4F, based on the trigonal bipyramid

A figure whose symmetry is D_{3h} can be distorted by destroying the 3-fold axis, giving C_{2v} symmetry; the σ_h, giving C_{3v} symmetry; or the three vertical planes of symmetry, giving D_3. Figures 3.32 and 3.33 illustrate these possibilities.

If the plane hexagon of D_{6h} symmetry is distorted so as to form a hexagonal based pyramid, the point group C_{6v}, which is not so far known as a molecular shape, is formed.

We can also have distortions of D_{6h} in which only 2-fold axes are retained;

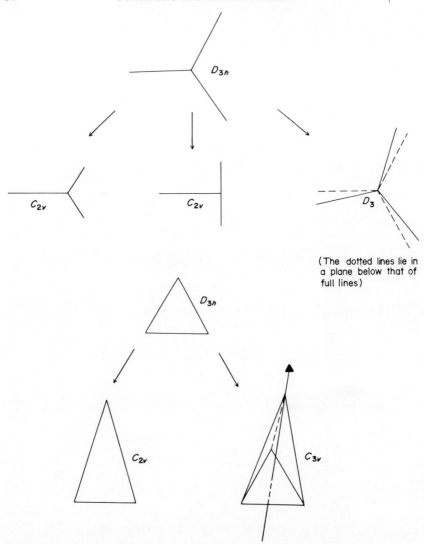

Figure 3.33   Figures with D_{3h} symmetry and their reduction to lower symmetries

Figure 3.14(a) shows how D_{6h} symmetry may be reduced to D_{2h}. In Figure 3.30, the figure of C_{6h} symmetry derived from that of D_{6h} symmetry may be reduced to one of C_{2h} symmetry by removing any two of its three arms.

The very low symmetry point groups C_2, C_s, and C_i can be obtained in so many ways that it is not worth considering them in detail. Although the above treatment is not exhaustive, it deals with the more important possibilities, and

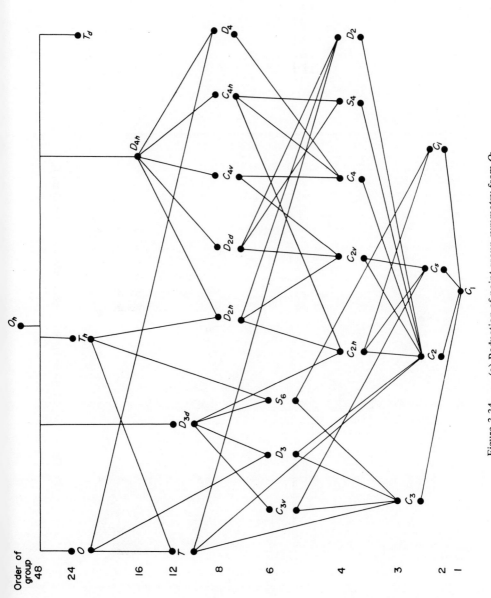

Figure 3.34   (a) Reduction of point group symmetry from O_h

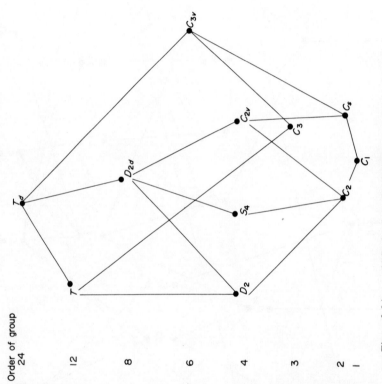

Figure 3.34   (b) Reduction of point group symmetry from T_d

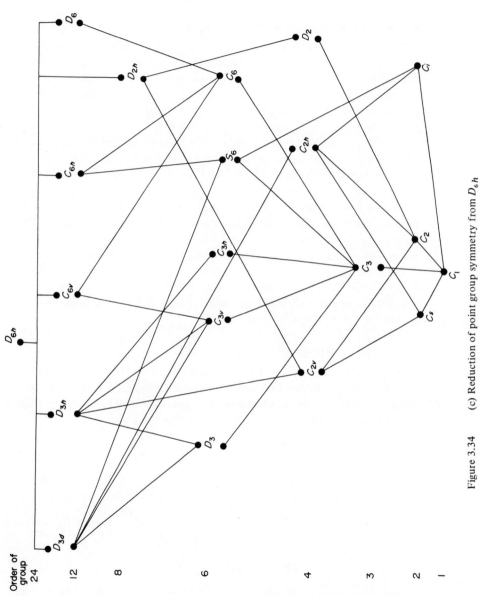

Figure 3.34 (c) Reduction of point group symmetry from D_{6h}

Figures 3.34 a, b, and c give a complete chart of the possible routes by which one point group can be formed from another.

Stereographic Projections and Point Groups

As we saw in Chapter 1 a stereographic projection on the equatorial circle of a sphere is a convenient way to represent three-dimensional symmetry in two-dimensions. Among the most useful of these projections are those which represent the point groups. In order to describe the 27 crystallographic axial point groups we need only provide a means of representing horizontal and vertical mirror planes and rotation axes (including rotor-reflection axes) on the equatorial circle of a sphere which contains the central point of the group at its centre. To describe the cubic point groups, however, we also need to be able to represent planes and axes which make angles other than 90° or 0° with the equatorial plane.

Two different types of diagram are useful in describing point group symmetry; these are the *Point-group Diagram* and the *Equivalent Positions Diagram*. The point group diagram is simply a projection of all of the symmetry elements of the group. The equivalent positions diagram also shows, in projection, what happens to any general position (x,y,z) on operation of the symmetry elements of the group. The symbols used to represent the various symmetry elements will be described as we gradually build up a set of stereographic projections to represent all 32 of the crystallographic point groups.

HORIZONTAL MIRROR PLANES

A horizontal mirror plane is a plane coincident with the equatorial circle which is used for the projection. In order to distinguish between point groups with and without a horizontal plane we enclose the projection in an open circle if there is no plane, and in a closed circle if there is a horizontal plane.

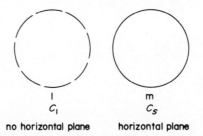

1   m
C_1 C_s

no horizontal plane   horizontal plane

The point group which contains no symmetry elements other than the identity is $1(C_1)$ and so the open circle is the point group diagram for this point group. Similarly, the point group which contains only a horizontal mirror plane is m (C_s) and the closed circle is the point group diagram for this point group.

The equivalent positions diagram for any point group is obtained by taking a projection of any general position (x,y,z) in the northern hemisphere of the sphere. Since this point is in the northern hemisphere it will be represented in projection as ●. If the operation of any symmetry element converts this point to a point in the southern hemisphere, it will then be represented in projection as ○. The equivalent positions diagrams for the point groups 1 and m are:

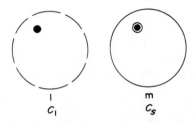

because in 1 (C_1) there are no symmetry elements to operate and in m (C_s) the only symmetry element reflects the position in the northern hemisphere into the southern hemisphere, i.e. it converts the general position (x,y,z) into $(x,y,-z)$. The example of a molecule with point group C_1 (1) from Table 3.5 is CHFClBr. If we take the centre of the hydrogen atom as a general position in this tetrahedral molecule, we can see that there is no other equivalent position in the molecule and it is this fact that the equivalent positions diagram represents. The molecule with point group C_s (m) from Table 3.5 is the bent planar molecule HOCl and again the equivalent positions diagram for this point group represents the fact that any position in the molecule must be reflected across the plane of the molecule.

VERTICAL ROTATION AXES AND ROTOR-REFLECTION AXES

These axes must cut the north and south poles of the sphere and will, on projection, lie at the centre of the equatorial circle. This can be represented by placing the appropriate polygon for the axis at the centre of the stereographic projection, i.e.

respectively for 2-, 3-, 4- and 6-fold rotation axes and

respectively for 2-, 3-, 4- and 6-fold rotor-reflection axes with their implied rotation axes where appropriate.

The point group diagrams for the groups which contain rotation axes as their only symmetry elements are thus:

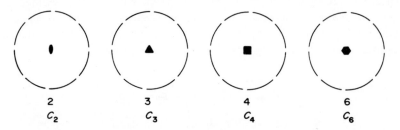

and the corresponding equivalent positions diagrams are:

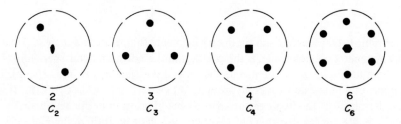

because the n-fold axes rotate the general position through $360°/n$ but keep it in the northern hemisphere. If we take the molecule H_2O_2 which has the point group C_2 (2), we can see from Figure 3.11 that any position (for example the centre of one of the hydrogen atoms) has an equivalent position (the centre of the other hydrogen atom) related to it by a 2-fold axis.

The point group diagrams for the groups which contain only rotor-reflection axes are:

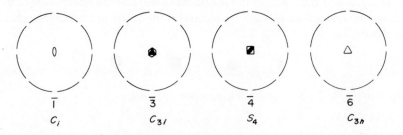

Remember that the Hermann-Mauguin symbols for these axes (groups) are those of the rotor-reflection axes, while the diagrammatic symbols are those for rotor-inversion. Because each rotation is combined with a reflection, the positions in the equivalent positions diagram are alternately in the northern and southern hemisphere. It is not possible to construct an equivalent positions diagram for

$\bar{6}$ and so the diagram for 3/m which is an identical symmetry to $\bar{6}$ is used. H_3BO_3 (Figure 3.18) has $3/m \equiv \bar{6}$ symmetry. The equivalent positions diagrams for these groups are:

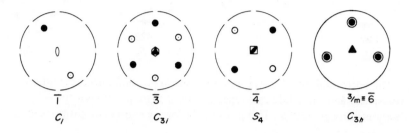

The implied 3-fold rotation axis in C_{3h} ($\bar{6}$) and the implied 2-fold axis in S_4 ($\bar{4}$) are clearly seen from these projections. The equivalent positions for S_4 ($\bar{4}$) can be verified for a real case by reference to the spiran molecule of Figure 3.12.

Apart from $3/m = \bar{6}$, the other point groups which contain an n-fold vertical axis and a horizontal mirror plane are 2/m, 4/m and 6/m, the equivalent positions diagrams for which are:

HORIZONTAL SYMMETRY AXES

Horizontal symmetry axes lie in the plane of the stereographic projection along one of the diagonals of the circle. They are denoted by placing the appropriate polygons for the symmetry axes at the points where the diagonal cuts the surface of the circle and drawing in the diagonal as a dashed line. For example, a 2-fold horizontal rotation axis would be denoted by

The point group diagrams for the groups which contain an n-fold vertical rotation axis and n 2-fold rotation axes are:

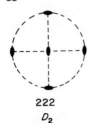   

| 222 | 32 | 422 | 622 |
| D_2 | D_3 | D_4 | D_6 |

The horizontal 2-fold axes rotate the general position into the southern hemisphere and the n-fold axis repeats both of these positions n times by rotations through $360°/n$. This gives the following equivalent positions diagrams.

   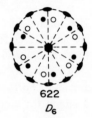

| 222 | 32 | 422 | 622 |
| D_2 | D_3 | D_4 | D_6 |

The hypothetical gauche forms of allene (Figure 3.13) and of ethane (Figure 3.15) have D_2 and D_3 symmetry respectively.

VERTICAL MIRROR PLANES

In projection, these planes are diagonals of the circle and are represented by drawing in the diagonal as a full line —————————. If a horizontal symmetry axis which is normally denoted by a dashed line, as shown above, also lies in a mirror plane, then the diagonal is drawn as a full line, e.g.

The equivalent positions diagrams for the groups which contain an n-fold rotation axis and n vertical mirror planes are:

 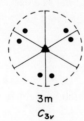 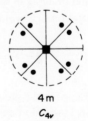 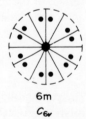

| mm | 3m | 4m | 6m |
| C_{2v} | C_{3v} | C_{4v} | C_{6v} |

The molecules CH_2Cl_2 and SF_4 (Figure 3.19) have C_{2v} symmetry.

The equivalent positions diagrams for the remaining six axial point groups are:

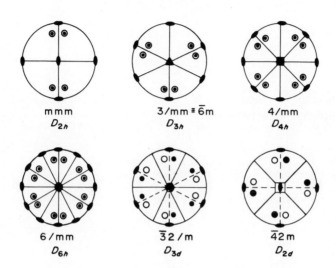

A sloping plane of symmetry cuts the sphere in a great circle. Projection of this
leads to curves lines (arcs of circles) in the equatorial circle. For example the
diagonal plane of a cube cuts the sphere in a great circle at the four corners of

and molecules having the symmetry of two of them, the S_6 sulphur allotrope
(Figures 3.16 and 3.17) (D_{3d}) and 1, 3, 5-trichlorobenzene (D_{3h}) have been
discussed earlier

SYMMETRY ELEMENTS NOT IN HORIZONTAL OR VERTICAL PLANES

An axis which cuts the plane of projection at an angle other than $0°$ or $90°$
is shown by placing the appropriate polygons at the points, in projection, where
the axis cuts the sphere. These points are then joined by a dotted line (no mirror
plane) or by a full line if the axis is coincident with a mirror plane. For example
a 3-fold axis of this type would be represented as:

the cube which lie in the plane. When this type of plane is projected we obtain a diagram such as:

The equivalent positions diagrams for the cubic point groups which contain axes and planes of these types are:

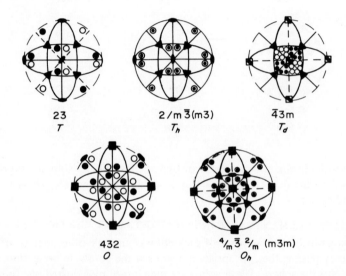

MULTIPLICATION TABLES

Stereographic projection provides a means of working out the components of a multiplication table. Figure 3.35 shows a proof, using stereographic projection, that $\sigma_{v1} \times C_3^1 \equiv \sigma_{v2}$ for the multiplication table for C_{3v} (Table 2.2).

PROBLEMS

1. Pick out the rotor-reflection axes in the following molecules:
    (a) ethane   (b) *trans*-1, 3-butadiene
    (c) BF_3 (d) SF_6

2. Work out which rotor-reflection axes are equivalent to the rotor-inversion axes $\bar{8}$ and $\overline{10}$.

3. Show that the following pairs of point groups are equivalent:
    (a) C_{5h} and S_5 (b) C_{5i} and S_{10}
    (c) C_{2i} and C_{2h} (d) C_{3h} and S_{3h}

Take the stereographic projection for C_{3v} with axes numbered as

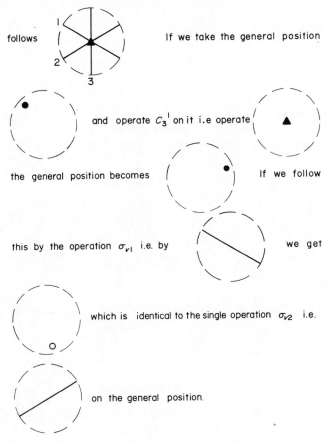

follows   If we take the general position

and operate $C_3{}^1$ on it i.e operate

the general position becomes   If we follow

this by the operation σ_{v1} i.e. by   we get

which is identical to the single operation σ_{v2} i.e.

on the general position.

Figure 3.35   Stereographic projection and multiplication tables

4. We do not designate any point group as being of the type S_{nv}. If a molecule has an S_{2n} axis where $n > 1$ together with n planes of symmetry which intersect in that axis, what other elements of symmetry must it have and how else could its point group be designated?

5. Why is the point group D_{4d} not a crystallographic point group?

6. What are the point groups of the capital letters of the alphabet, assuming that they have their most symmetrical forms. Assume also that they are solid figures.

7. What are the point groups of the following forms of the capital letters and numbers:

(a) I as distinct from I
(b) L as distinct from L where both arms are of the same length

4 – Coordination

5 – Coordination

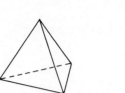

Tetrahedron

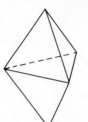

Trigonal – biprism

Square – based
prism

6 – Coordination

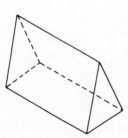

Trigonal prism

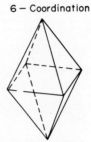

Octahedron

Pentagonal –
based
pyramid

7 – Coordination

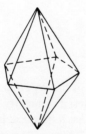

Pentagonal biprism

Hexagonal –
based
pyramid

8 – Coordination

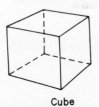

Cube

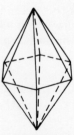

Hexagonal biprism

Figure 3.36

(c)   X as distinct from X where all angles are 90°
(d)   4 as distinct from 4
(e)   3 as distinct from 3
(f)   8 as distinct from 8?

8. What are the point groups of the solid figures shown in Figure 3.36, which represent possible 4-, 5-, 6-, 7- and 8- co-ordination polyhedra.

9. What are the point groups of the following molecular shapes?

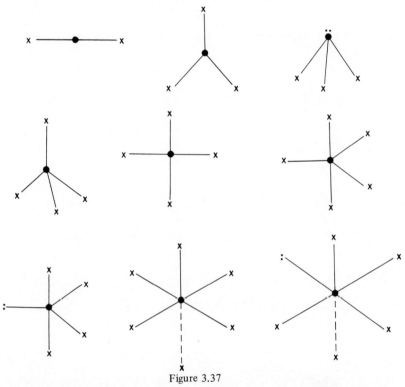

Figure 3.37

10. What are the point groups of the organic molecules shown in Figure 3.38?

11. What are the point groups of all the possible configurations obtained by replacing one X by a Y in each of the molecular shapes of Figure 3.37?

12. What are the point groups of all of the possible configurations obtained by replacing two of the X groups by two Y groups in each of the molecular shapes of Figure 3.37?

13. Draw the stereographic projections for C_{5v}, D_7, C_{5h}.

14. Show as a stereogram what happens to a general point on operation of the following symmetry elements:

(a) σ_h, (b) σ_v, (c) C_2, (d) C_3^2, (e) S_4^1, (f) S_2, (g) S_6^5.

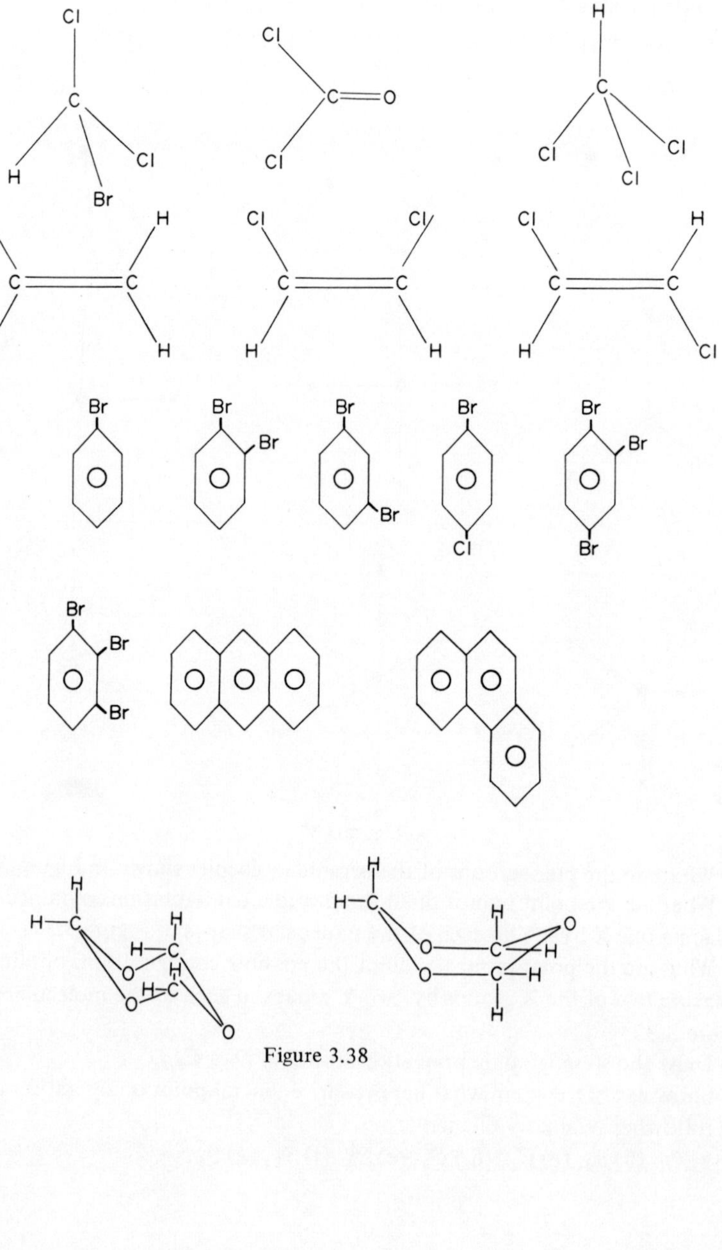
Cl
C
H
Cl
Br
Cl
C
O
Cl
H
C
Cl
Cl
Cl
Cl
H
C
C
Cl
H
Cl
Cl
C
C
H
H
Cl
H
C
C
H
Cl
Br
Br
Br
Br
Br
Br
Br
Cl
Br
Br
Br
Br
Br
Br
H
H
C
H
C
H
C
O
O
O
H
H
C
H
C
O
O
O
C
H
C
H
H

Figure 3.38

4

Space Group Symmetry

In Chapter 3 we considered all of the possible symmetries arising from simple and compound elements of rotation and reflection. We saw that, if we restrict the order of the rotation or rotor-reflection axes present to $n = 1, 2, 3, 4$ or 6 then there are only 32 possible overall symmetries; these are the 32 crystal point groups. The limitation on the order of the axes arises from the need to pack asymmetric groups in space to form crystals obeying the requirements of translation symmetry. If we consider the two dimensional analogy of laying tiles on a floor, it is clearly possible to cover all of the floor area with square, oblong, triangular or diamond shaped tiles. It is not, however, possible to cover all of the area with, for example, pentagonal or heptagonal tiles. Extending the analogy to three dimensions it is possible to close-pack units based on oblique, diamond, rectangular, hexagonal or square cells but we cannot form a close-packed crystal if the order of the axis is 5 or 7.

In this chapter we shall consider the possible symmetries arising from the presence of combinations of simple and compound elements of translation, rotation and reflection. We shall see that there are only 230 possible overall symmetries and these are the *230 space groups*.

Lattices and Cells

We saw in Chapter 2 that translation symmetry gives rise to plane and space arrays of asymmetric objects. These asymmetric objects can range from a single atom in some element crystals to large and complicated molecules in protein structures. The complexity of the asymmetric unit is not, however, important in discussing translation symmetry. Crystallographers simplify the plane and space arrays by replacing the asymmetric objects by points such that the environment of every point is the same. These points are called lattice points and the plane and space arrays become plane and space lattices. If we replace each seven in Figures 2.1 and 2.2 by points such that the environment of each point is the

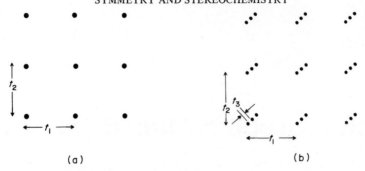

Figure 4.1   (a) Plane lattice (b) Space lattice

same (for example the point at which the two arms of the seven meet) then we obtain the plane and space lattices of Figure 4.1.

A plane lattice can be described by a number of sets of translations. In Figure 4.2 each of the sets of translations $t_1 t_2$, $t_1 t_3$, $t_1 t_4$, $t_4 t_5$, $t_5 t_6$ describe small cells which, when repeated in two-dimensional space completely describe the lattice. There is obviously an infinite number of such cells. $t_1 t_2$ and $t_1 t_4$ are called *Primitive Cells* because they contain within their boundaries only one lattice point. $t_1 t_3$, $t_4 t_5$ and $t_5 t_6$ are *Multiple Cells* because they contain more than one lattice point. Clearly any plane lattice can be described by any one of a large number of cells defined by two translations and the angle between them. The important cell, however, is the *Unit Cell* which is generally defined as the smallest or most symmetrical cell which on repetition in space will completely describe the lattice. The unit cell may be primitive or multiple depending upon the type of lattice. The five possible plane cells are listed in Table 4.1 and described in Figure 4.3.

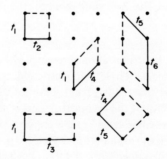

Figure 4.2   Sets of translations describing a plane lattice

Table 4.1. The five types of plane lattice

Cell Shape	Translations a and b	Angle γ	Figure
Oblique	$a \neq b$	$\gamma \neq 90°$	4.3(a)
Oblique (diamond)	$a = b$	$\gamma \neq 90$	4.3(b)
	The diamond cell is the smallest cell but it is often replaced by the centred cell of Figure 4.3(c) which, although larger, is a better description of the lattice.		
Rectangular	$a \neq b$	$\gamma = 90°$	4.3(d)
Square	$a = b$	$\gamma = 90°$	4.3(e)
Hexagonal	$a = b$	$\gamma = 60°$	4.3(f)
	or $a = b$	$\gamma = 120°$	4.3(g)
	This is a special case of a diamond cell		

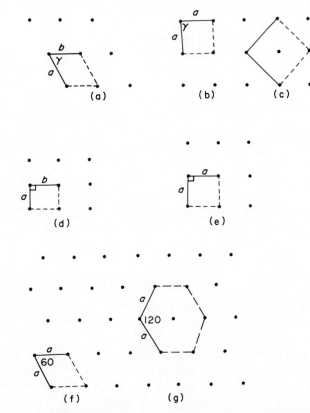

Figure 4.3   The five plane lattice types. (a) oblique, (b) and (c) the primitive and centred oblique (diamond), (d) oblong, (e) square, (f) and (g) primitive and centred hexagonal

Oblique and rectangular cells have 2-fold rotational symmetry, square cells have 4-fold rotational symmetry and hexagonal cells have 6-fold (cell with $60°$ angle) or 3-fold (cell with $120°$ angle) rotational symmetry.

In three dimensions the unit cell is described by three translations and three angles.

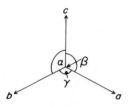

There are six possible shapes for the three-dimensional unit cells. These are the six *Crystal Classes* listed in Table 4.2.

Table 4.2. The six crystal classes

	Cell translations	Cell angles
Triclinic	$a \neq b \neq c$	$\alpha \neq \beta \neq \gamma$
Monoclinic	$a \neq b \neq c$	$\alpha = \gamma = 90°, \beta \neq 90°$
Orthorhombic	$a \neq b \neq c$	$\alpha = \beta = \gamma = 90°$
Hexagonal	$a = b \neq c$	$\alpha = \beta = 90° \ \gamma = 120°$
Rhombohedral	$a = b = c$	$\alpha = \beta = \gamma \neq 90°$
Tetragonal	$a = b \neq c$	$\alpha = \beta = \gamma = 90°$
Cubic	$a = b = c$	$\alpha = \beta = \gamma = 90°$

The hexagonal and rhombohedral cells belong to the same crystal class and are based on the alternative 6-fold and 3-fold cells of the hexagonal plane lattice.

Every crystal class must have a primitive cell associated with it but multiple unit cells are also possible. The four possible types of cell, primitive (P), body-centred (I), face-centred all on faces (F) and face-centred on one face (A, B or C depending upon the face which is centred) are shown in Figure 4.4. Primitive cells must of course contain only one lattice point within their boundaries; the body- and face-centred (one face) cells contain two lattice points and the face-centred (all faces) cell contains four lattice points.

Primitive cells exist in all six crystal classes and we can work out which of the multiple cells are possible for each crystal class. If we start with a primitive lattice cell we can add additional lattice points to it in the following three ways:

(a) at the centre of the cell to make it body-centred
(b) at the centres of all six faces to make it face-centred on all faces
(c) at the centres of one pair of opposite faces to make it face-centred on one face.

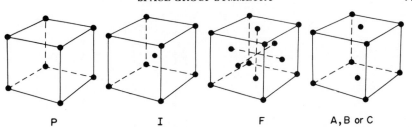

P   I   F   A, B or C

Figure 4.4   The four possible three dimensional unit cells, Primitive (P), Body-centred (I), Face-centred on all faces (F), Face-centred on one face (A, B or C). An atom at a corner of a unit cell contributes $\frac{1}{8}$ to that cell, at an edge $\frac{1}{4}$ and on a face $\frac{1}{2}$

Having done this for each of the crystal classes we can check which of the forms of centring are possible by answering the following three questions:

1. Is the new array of points still a lattice? If the new array cannot be described completely by a set of three translations and angles, the centring is not possible for the crystal class being considered.

2. Has the symmetry of the cell been altered? Obviously if any type of centring alters the symmetry of the cell, it no longer belongs to the crystal class being considered.

3. If the arrangement is still a lattice and the symmetry has not altered, is the new arrangement identical to any of the other arrangements for the crystal class being considered? This allows for the possibility that one of the centred cells may have an identical symmetry to the primitive cell or to one of the other centred cells.

In the cubic system the body-centred and face-centred (on all faces) cells form lattices, remain cubic and are not identical with each other or with the primitive cell. Thus P, I and F cells are permitted in the cubic system. Addition of lattice points to one pair of opposite faces of a primitive cubic cell destroys the 4-fold axes parallel to these faces and the 3-fold axes across the cube diagonals and, therefore, the cubic symmetry (see Figure 4.5 a). Single face-centred lattices (e.g. C) are thus not allowed in cubic symmetry.

In the tetragonal system the body-centred cell is permitted. The lattice based on a tetragonal cell centred on all faces can also be described in terms of a tetragonal body-centred cell, i.e. the cell at 45° to the centred lattice in Figure 4.5 b. The relative lengths of the translations describing these two cells are not the same but nevertheless they have the same symmetry. Thus for tetragonal systems F = I. By similar arguments it can be shown that face-centring on the oblong faces (A and B) destroys the tetragonal symmetry (i.e. the 4-fold axis) but centring on the C face produces a lattice which can also be represented by a primitive cell at 45° to the original. Thus for the tetragonal system C = P.

In the orthorhombic system all four possible cells are permitted.

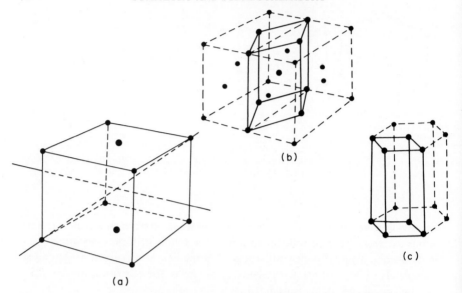

Figure 4.5   (a) Face-centring of cubic cell on one face destroys cubic symmetry.
                 The cube diagonals are no longer 3-fold axes and the axes through the
                 non-centred faces are now 2-fold not 4-fold
                 (b) The equivalence of face-centred (F) dotted lines, and body-centred (I)
                 full lines, tetragonal cells
                 (c) The primitive and C-centred hexagonal cells

   In the hexagonal system I, F, A and B centring are not permitted. The
simplest cell is primitive but the lattice is often described in terms of the larger
C-centred cell shown in Figure 4.5 c. The only cell allowed in the rhombohedral
system is the primitive cell which is usually given the symbol R. Again in the
rhombohedral system it is possible to pick out a larger centred cell which is
sometimes a better description of the lattice.
   In the monoclinic system B-centring (i.e. centring on the unique face) is
equivalent to the primitive cell. A-centring and the centred cells I and F are
equivalent to C. The primitive cell is the only allowed triclinic cell.
   There is thus a total of fourteen possible types of three-dimensional cell.
These are the fourteen *Bravais lattices* shown in Figure 4.6.
   The lattice type of a crystal has a marked effect on its X-ray diffraction
patterns and, as we shall see in Chapter 6, it is relatively simple to obtain information on the size and shape of a unit cell and on its type of Bravais lattice from
single crystal X-ray data.

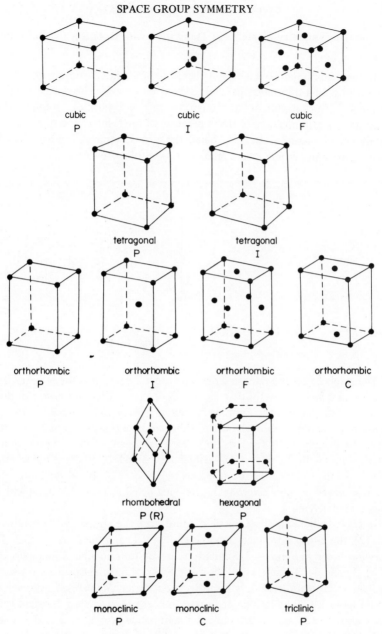

Figure 4.6   The 14 Bravais lattices

Compound Symmetry Elements of Translation with Rotation and Reflection

SCREW AXES

In these compound symmetry elements a right-handed rotation of $360°/n$ is combined with a translation parallel to the rotation axis. This is illustrated in Figure 4.7 by a rotation of 7a through $\phi = 360°/n$ to position 7b followed by a translation, t, to position 7c. The operation of the symmetry element is equivalent to a screw motion and hence the term screw axis. In Figure 4.7 the asymmetric units 7a and 7c are related by an n-fold screw axis.

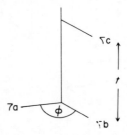

Figure 4.7   An n-fold screw axis

The symbols used to denote screw axes are n_m where n is the order of the rotation axis and m denotes that the length of the translation t is m/n of the unit cell edge. For example the symbols 2_1 denote a screw axis formed by rotation of an asymmetric object through $360°/2$ followed by a translation of $1/2$ of the cell edge. The diagrammatic symbols for screw axes consist of the solid polygon for the rotation axes with the sides of the polygon extended to represent the translations m/n. Since the order of the rotation axis is restricted in crystal symmetry there are only eleven possible screw axes. These are listed in Table 4.3 along with their diagrammatic symbols.

In crystals there are three possible 2-fold axes, the normal 2-fold rotation axis (2), the 2-fold rotor-inversion axis ($2 \equiv m \equiv \bar{1}$) and the 2_1 screw axis. These axes are shown in Figure 4.8. Similarly there are four possible 3-fold axes: 3, $\bar{3}$ ($\equiv \bar{6}$), 3_1 and 3_2. These are also shown in Figure 4.8. In discussing screw axes we have so far considered only right-handed rotation. We can see, however, from Figure 4.8 that a left-handed rotation does not produce a different type of screw axis. The 3_1 axis for example is a right-handed rotation of $360°/3$ combined with a translation of $1/3$ of the cell edge, while 3_2 is a right-handed rotation of $360°/3$ combined with a translation of $2/3$ of the cell edge. The 3_2 axis can, however, also be formed by a left-handed rotation of $360°/3$ followed by a translation of $1/3$ of the cell edge, and 3_1 by the left-handed rotation followed by a $2/3$

Table 4.3   The Eleven screw axes

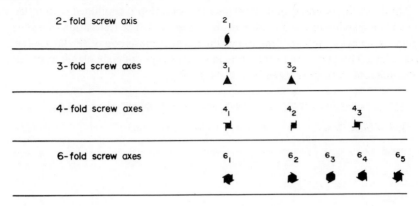

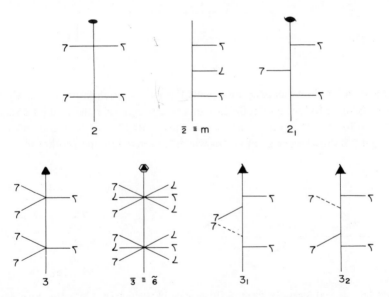

Figure 4.8   The possible types of 2-fold and 3-fold axes in crystal symmetry

translation. This means that 3_1 and 3_2 are in fact mirror images of one another. This can be seen from Figure 4.8 and from the symbols in Table 4.3. By similar arguments 4_3 is the mirror image of 4_1, 6_5 is the mirror image of 6_1 and 6_4 the mirror image of 6_2. The diagrammatic symbols of Table 4.3 show these relationships.

Again the presence of screw axes can affect the X-ray diffraction pattern of a crystal and this will be discussed in Chapter 6.

GLIDE PLANES

In these compound symmetry elements translation is combined with reflection in a plane parallel to the direction of the translation. The translation of a glide plane is always an integral fraction of the normal translation of the lattice (i.e. the cell edge). This gives rise to the three important types of glide plane: axial, diagonal and diamond which are listed in Table 4.4

Table 4.4. Types of glide plane for a crystal with unit cell edges a,b,c

Type of Glide Plane	Extent of the glide (translation)	Symbol
axial	$a/2$	a
axial	$b/2$	b
axial	$c/2$	c
diagonal	$\dfrac{a+b}{2}$, or $\dfrac{b+c}{2}$, or $\dfrac{a+c}{2}$	n
diamond	$\dfrac{a+b}{4}$, or $\dfrac{b+c}{4}$, or $\dfrac{a+c}{4}$	d

Figure 4.9 a shows an axial glide plane formed by a translation of $a/2$ which moves 7a to 7b followed by reflection in a mirror plane parallel to the translation direction to 7c. The figures 7a and 7c are thus related by the axial glide plane. Figure 4.9 b shows an array of sevens which are related to one another by a

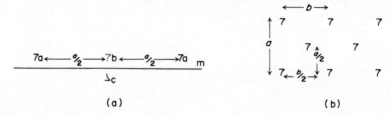

(a)   (b)

Figure 4.9   (a) An axial glide plane (b) an array of sevens related by a diagonal glide plane

diagonal glide plane with a glide of $\dfrac{a+b}{2}$. Again this type of symmetry element can be detected from single crystal X-ray diffraction patterns as described in Chapter 6.

Space Groups

We are now in a position to describe the combinations of translation, rotation and reflection symmetry which make up the 230 space groups. First of all, however, let us consider the information that we have available to describe three-dimensional symmetry.

1. We know that there are 32 crystal point groups, i.e. combinations of rotation, reflection and rotor-reflection symmetry. Any asymmetric unit which is built into a space lattice must have the symmetry of one of these point groups.

2. We know that there are six crystal classes; triclinic with 1-fold symmetry, monoclinic and orthorhombic with 2-fold symmetry, tetragonal which has a 4-fold axis, cubic which contains both 3-fold and 4-fold or 2-fold axes and hexagonal and rhombohedral with 6-fold and 3-fold axes respectively.

3. We know that, for some of the crystal classes, certain centred cells are permissible in addition to the primitive cell.

4. We know that, in addition to rotation and rotor-reflection axes, we have the possibility of screw axes in crystals and that reflection planes in molecular symmetry can replaced by glide planes in crystals.

The 230 space groups are obtained by considering the application of point group symmetry to the crystal classes and the Bravais lattices and allowing for the compound translation symmetry elements of screw axes and glide planes. If we work through a number of examples of combining all of the possible symmetry elements, it should become obvious how the 230 space groups have been built up.

First of all we must consider the application of the 32 point groups to the crystal classes. Not every point group is compatible with any crystal class. For example, for an asymmetric unit to fit into a cubic unit cell it must itself have cubic symmetry. The only point groups which are compatible with the cubic crystal class are thus the five cubic point groups. Similarly the requirement of a 4-fold axis for tetragonal crystals, of a 6-fold axis for hexagonal crystals and of a 3-fold axis for rhombohedral lattices limits the number of possible point groups. Table 4.5 lists the point groups compatible with each of the crystal classes.

Table 4.5. The point groups compatible with the crystal classes

Triclinic	1, $\bar{1}$
Monoclinic	2, m, 2/m
Orthorhombic	222, mm2, mmm
Tetragonal	4, $\bar{4}$, 4/m, 422, 4mm, $\bar{4}$2m, 4/mmm
Rhombohedral	3, $\bar{3}$, 32, 3m, $\bar{3}$m
Hexagonal	6, $\bar{6}$, 6/m, 622, 6mm, $\bar{6}$m2, 6/mmm
Cubic	23, m3, 432, $\bar{4}$3m, m3m

The next stage in the development of the space groups is to consider the application of the point groups to every possible Bravais lattice of the crystal class. The triclinic system is a simple one because it has only one lattice type, the primitive lattice, and there are two point groups compatible with this three-dimensional symmetry (i.e. 1 and $\bar{1}$). There are no rotation axes or mirror planes in the point groups 1 and $\bar{1}$ which could be replaced by the alternative screw axes or glide planes and, therefore, there are only two triclinic space groups:

(i) combination of the point group 1 with the primitive lattice P, i.e. P1

(ii) combination of the point group $\bar{1}$ with the primitive lattice P, i.e. P$\bar{1}$.

In the monoclinic system there are two possible Bravais lattices P and C and three point groups. To have the centring on the C face, the monoclinic cell must be described in a form with b as the unique axis, that is with $a \neq b \neq c$, $\alpha = \gamma = 90°$, $\beta \neq 90°$. We can combine the primitive cell with the three point groups to give the space groups P2, Pm and P2/m. In monoclinic crystal symmetry we can have the screw axis 2_1 as an alternative to 2 and the glide plane c as an alternative to m (a,b,n and d glide planes are not compatible with monoclinic symmetry). These alternatives give rise to space groups with all the possible combinations, P2_1, Pc, P2_1/m, P2_1/c. We can also combine the C-centred cell with the three point groups to give C2, Cm and C2/m. The same alternatives are possible but only two of them (Cc and C2/c) are unique space groups — the other combinations with the alternative compound translation elements have the same symmetry as one of the other groups in the set. The application of the three monoclinic point groups to the two possible Bravais lattices of this system therefore gives rise to a total of 13 space groups. Table 4.6 shows the allocation of the space groups to the Bravais lattices in each crystal class.

Table 4.6. Distribution of space groups among the Bravais Lattices

	No. of point Groups	No. of Space Groups which are:				Total No. of Space Groups
		P	F	I	C(or A)	
Triclinic	2	2	–	–	–	2
Monoclinic	3	8	–	–	5	13
Orthorhombic	3	30	5	9	15	59
Tetragonal	7	49	–	19		68
Rhombohedral	5	25*	–	–	–	25
Hexagonal	7	27*	–	–	–	27
Cubic	5	15	11	10		36

* The smallest hexagonal and rhombohedral cells are P but larger centred cells are sometimes taken as a better description of these systems.

Taking one further example from the tetragonal system, it can be shown that six space groups arise from combinations of the point group 4 with the two possible Bravais lattices of this system (P and I). These space groups are:

P combined with the 4-fold rotation axis   P4

P combined with a 4_1 screw axis   P4_1

P combined with a 4_2 screw axis   P4_2

P combined with a 4_3 screw axis $P4_3$
I combined with the 4-fold rotation axis $I4$
I combined with a 4_1 screw axis $I4_1$

All other combinations reduce to one of this set.

SPACE GROUP NOMENCLATURE

We have already written down the symbols for a number of space groups. P1, P$\bar{1}$, Pm, C2/m, P4, I4, etc. are all Hermann-Mauguin symbols for space groups. As with point-group symmetry, the logic of the Hermann-Mauguin nomenclature is the listing of the number of symmetry elements required to describe the space groups and the rules for listing these symbols are set out in this chapter. Space groups can also be described in terms of a Schoenflies notation which is an extension of the point group symbolism. Unlike the Hermann-Mauguin nomen-clature, however, the main crystallographic symmetry elements are not immediately clear from the Schoenflies notation. For this reason the Schoenflies symbols are given in the complete list of space groups (Table 4.8) but are not discussed in detail in this book.

The first Hermann-Mauguin symbol for a space group is always that of the lattice type, i.e. P,I,F,C,A. The subsequent symbols then refer to the nature of the axes (rotation, rotor-inversion or screw) corresponding to definite directions in the lattice. Symbols representing the symmetries must be given for enough directions in the lattice to provide a complete description of the space symmetry. The symbols which are used are those which we have already defined and their meanings in space group nomenclature are given in Table 4.8 with reference to a particular direction in space (the a axis of the lattice). The symbols have similar meanings with reference to other directions in space as can be seen from the examples of space group nomenclature quoted later in this chapter.

Table 4.7. Meanings of typical symbols used in space group nomenclature given with reference to the a axis

2	the a axis is 2-fold rotation axis
2_1	the a axis is a 2-fold screw axis
$\bar{4}$	the a axis is a 4-fold rotor-inversion axis
m	the a axis is a 2-fold rotor-inversion axis, i.e. there is a mirror plane normal to it
b	there is an axial glide plane perpendicular to a with glide $b/2$
c	there is an axial glide plane perpendicular to a with glide $c/2$
n	there is a diagonal glide plane perpendicular to a with glide $\dfrac{b+c}{2}$
d	there is a diamond glide plane perpendicular to a with glide $\dfrac{b+c}{4}$

SYMMETRY AND STEREOCHEMISTRY

Table 4.8. The 230 Space Groups

H.M. (short) = short form of the Hermann-Mauguin symbols
H.M. (full) = full form of the Hermann-Mauguin symbols
S = Schoenflies symbols

Space Group No.	H.M. (short)	H.M. (full)	S
Triclinic			
1	P1	P1	C_1^1
2	P$\bar{1}$	P$\bar{1}$	C_i^1
Monoclinic			
3	P2	P121	C_2^1
4	P2$_1$	P12$_1$1	C_2^2
5	C2	C121	C_2^3
6	Pm	P1m1	C_s^1
7	Pc	P1c1	C_s^2
8	Cm	C1m1	C_s^3
9	Cc	C1c1	C_s^4
10	P2/m	P12/m1	C_{2h}^1
11	P2$_1$/m	P12$_1$/m1	C_{2h}^2
12	C2/m	C12/m1	C_{2h}^3
13	P2/c	P12/c1	C_{2h}^4
14	P2$_1$/c	P12$_1$/c1	C_{2h}^5
15	C2/c	C12/c1	C_{2h}^6
Orthorhombic			
16	P222	P222	D_2^1
17	P222$_1$	P222$_1$	D_2^2
18	P2$_1$2$_1$2	P2$_1$2$_1$2	D_2^3
19	P2$_1$2$_1$2$_1$	P2$_1$2$_1$2$_1$	D_2^4
20	C222$_1$	C222$_1$	D_2^5
21	C222	C222	D_2^6
22	F222	F222	D_2^7
23	I222	I222	D_2^8
24	I2$_1$2$_1$2$_1$	I2$_1$2$_1$2$_1$	D_2^9
25	Pmm2	Pmm2	C_{2v}^1
26	Pmc2$_1$	Pmc2$_1$	C_{2v}^2
27	Pcc2	Pcc2	C_{2v}^3
28	Pma2	Pma2	C_{2v}^4
29	Pca2$_1$	Pca2$_1$	C_{2v}^5
30	Pnc2	Pnc2	C_{2v}^6
31	Pmn2$_1$	Pmn2$_1$	C_{2v}^7
32	Pba2	Pba2	C_{2v}^8
33	Pna2$_1$	Pna2$_1$	C_{2v}^9
34	Pnn2	Pnn2	C_{2v}^{10}
35	Cmm2	Cmm2	C_{2v}^{11}
36	Cmc2$_1$	Cmc2$_1$	C_{2v}^{12}
37	Ccc2	Ccc2	C_{2v}^{13}
38	Amm2	Amm2	C_{2v}^{14}
39	Abm2	Abm2	C_{2v}^{15}
40	Ama2	Ama2	C_{2v}^{16}
41	Aba2	Aba2	C_{2v}^{17}
42	Fmm2	Fmm2	C_{2v}^{18}

Space Group No.	H.M. short	H.M. (full)	S
43	Fdd2	Fdd2	C_{2v}^{19}
44	Imm2	Imm2	C_{2v}^{20}
45	Iba2	Iba2	C_{2v}^{21}
46	Ima2	Ima2	C_{2v}^{22}
47	Pmmm	$P2/m2/m2/m$	D_{2h}^{1}
48	Pnnn	$P2/n2/n2/n$	D_{2h}^{2}
49	Pccm	$P2/c2/c2/m$	D_{2h}^{3}
50	Pban	$P2/b2/a2/n$	D_{2h}^{4}
51	Pmma	$P2_1/m2m/2/a$	D_{2h}^{5}
52	Pnna	$P2/n2_1/n2/a$	D_{2h}^{6}
53	Pmna	$P2/m2/n2_1/a$	D_{2h}^{7}
54	Pcca	$P2_1/c2/c2/a$	D_{2h}^{8}
55	Pbam	$P2_1/b2_1/a2/m$	D_{2h}^{9}
56	Pccn	$P2_1/c2_1/c2/n$	D_{2h}^{10}
57	Pbcm	$P2/b2_1/c2_1/m$	D_{2h}^{11}
58	Pnnm	$P2_1/n2_1/n2/m$	D_{2h}^{12}
59	Pmmn	$P2_1/m2_1/m2/n$	D_{2h}^{13}
60	Pbcn	$P2_1/b2/c2_1/n$	D_{2h}^{14}
61	Pbca	$P2_1/b2_1/c2_1/a$	D_{2h}^{15}
62	Pnma	$P2_1/n2_1/m2_1/a$	D_{2h}^{16}
63	Cmcm	$C2/m2/c2_1/m$	D_{2h}^{17}
64	Cmca	$C2/m2/c2_1/a$	D_{2h}^{18}
65	Cmmm	$C2/m2/m2/m$	D_{2h}^{19}
66	Cccm	$C2/c2/c2/m$	D_{2h}^{20}
67	Cmma	$C2/m2/m2/a$	D_{2h}^{21}
68	Ccca	$C2/c2/c2/a$	D_{2h}^{22}
69	Fmmm	$F2/m2/m2/m$	D_{2h}^{23}
70	Fddd	$F2/d2/d2/d$	D_{2h}^{24}
71	Immm	$I2/m2/m2/m$	D_{2h}^{25}
72	Ibam	$I2/b2/a2/m$	D_{2h}^{26}
73	Ibca	$I2/b2/c2/a$	D_{2h}^{27}
74	Imma	$I2/m2/m2/a$	D_{2h}^{28}

Tetragonal

75	P4	P4	C_4^1
76	$P4_1$	$P4_1$	C_4^2
77	$P4_2$	$P4_2$	C_4^3
78	$P4_3$	$P4_3$	C_4^4
79	I4	I4	C_4^5
80	$I4_1$	$I4_1$	C_4^6
81	$P\bar{4}$	$P\bar{4}$	S_4^1
82	$I\bar{4}$	$I\bar{4}$	S_4^2
83	P4/m	P4/m	C_{4h}^1
84	$P4_2/m$	$P4_2/m$	C_{4h}^2
85	P4/n	P4/n	C_{4h}^3
86	$P4_2/n$	$P4_2/n$	C_{4h}^4
87	I4/m	I4/m	C_{4h}^5
88	$I4_1/a$	$I4_1/a$	C_{4h}^6
89	P422	P422	D_4^1
90	$P42_12$	$P42_12$	D_4^2
91	$P4_122$	$P4_122$	D_4^3
92	$P4_12_12$	$P4_12_12$	D_4^4

Space Group No.	H.M. (short)	H.M. (full)	S
93	$P4_2 22$	$P4_2 22$	D_4^5
94	$P4_2 2_1 2$	$P4_2 2_1 2$	D_4^6
95	$P4_3 22$	$P4_3 22$	D_4^7
96	$P4_3 2_1 2$	$P4_3 2_1 2$	D_4^8
97	$I422$	$I422$	D_4^9
98	$I4_1 22$	$I4_1 22$	D_4^{10}
99	$P4mm$	$P4mm$	C_{4v}^1
100	$P4bm$	$P4bm$	C_{4v}^2
101	$P4_2 cm$	$P4_2 cm$	C_{4v}^3
102	$P4_2 nm$	$P4_2 nm$	C_{4v}^4
103	$P4cc$	$P4cc$	C_{4v}^5
104	$P4nc$	$P4nc$	C_{4v}^6
105	$P4_2 mc$	$P4_2 mc$	C_{4v}^7
106	$P4_2 bc$	$P4_2 bc$	C_{4v}^8
107	$I4mm$	$I4mm$	C_{4v}^9
108	$I4cm$	$I4cm$	C_{4v}^{10}
109	$I4_1 md$	$I4_1 md$	C_{4v}^{11}
110	$I4_1 cd$	$I4_1 cd$	C_{4v}^{12}
111	$P\bar{4}2m$	$P\bar{4}2m$	D_{2d}^1
112	$P\bar{4}2c$	$P\bar{4}2c$	D_{2d}^2
113	$P\bar{4}2_1 m$	$P\bar{4}2_1 m$	D_{2d}^3
114	$P\bar{4}2_1 c$	$P\bar{4}2_1 c$	D_{2d}^4
115	$P\bar{4}m2$	$P\bar{4}m2$	D_{2d}^5
116	$P\bar{4}c2$	$P\bar{4}c2$	D_{2d}^6
117	$P\bar{4}b2$	$P\bar{4}b2$	D_{2d}^7
118	$P\bar{4}n2$	$P\bar{4}n2$	D_{2d}^8
119	$I\bar{4}m2$	$I\bar{4}m2$	D_{2d}^9
120	$I\bar{4}c2$	$I\bar{4}c2$	D_{2d}^{10}
121	$I\bar{4}2m$	$I\bar{4}2m$	D_{2d}^{11}
122	$I\bar{4}2d$	$I\bar{4}2d$	D_{2d}^{12}
123	$P4/mmm$	$P4/m2/m2m$	D_{4h}^1
124	$P4/mcc$	$P4/m2/c2/c$	D_{4h}^2
125	$P4/nbm$	$P4/n2/b2/m$	D_{4h}^3
126	$P4/nnc$	$P4/n2/n2/c$	D_{4h}^4
127	$P4/mbm$	$P4/m2_1/b2/m$	D_{4h}^5
128	$P4/mnc$	$P4/m2_1/n2/c$	D_{4h}^6
129	$P4/nmm$	$P4/n2_1/m2/m$	D_{4h}^7
130	$P4/ncc$	$P4/n2_1/c2/c$	D_{4h}^8
131	$P4_2 /mmc$	$P4_2/m2/m2/c$	D_{4h}^9
132	$P4_2 /mcm$	$P4_2/m2/c2/m$	D_{4h}^{10}
133	$P4_2 /nbc$	$P4_2/n2/b2/c$	D_{4h}^{11}
134	$P4_2 /nnm$	$P4_2/n2/n2/m$	D_{4h}^{12}
135	$P4_2 /mbc$	$P4_2/m2_1/b2/c$	D_{4h}^{13}
136	$P4_2 /mnm$	$P4_2/m2_1/n2/m$	D_{4h}^{14}
137	$P4_2 /nmc$	$P4_2/n2_1/m2/c$	D_{4h}^{15}
138	$P4_2 /ncm$	$P4_2/n2_1/c2/m$	D_{4h}^{16}
139	$I4/mmm$	$I4/m2/m2/m$	D_{4h}^{17}
140	$I4/mcm$	$I4/m2/c2/m$	D_{4h}^{18}
141	$I\,4_1 /amd$	$I4_1/a2/m2/d$	D_{4h}^{19}
142	$I\,4_1 /acd$	$I4_1/a2/c2/d$	D_{4h}^{20}

Rhombohedral

143	$P3$	$P3$	C_3^1

Space Group No.	H.M. (short)	H.M. (full)	S
144	$P3_1$	$P3_1^{\backslash}$	C_3^2
145	$P3_2$	$P3_2$	C_3^3
146	$R3$	$R3$	C_3^4
147	$P\bar{3}$	$P\bar{3}$	C_{3i}^1
148	$R\bar{3}$	$R\bar{3}$	C_{3i}^2
149	$P312$	$P312$	D_3^1
150	$P321$	$P321$	D_3^2
151	$P3_112$	$P3_112$	D_3^3
152	$P3_121$	$P3_121$	D_3^4
153	$P3_212$	$P3_212$	D_3^5
154	$P3_221$	$P3_221$	D_3^6
155	$R32$	$R32$	D_3^7
156	$P3m1$	$P3m1$	C_{3v}^1
157	$P31m$	$P31m$	C_{3v}^2
158	$P3c1$	$P3c1$	C_{3v}^3
159	$P31c$	$P31c$	C_{3v}^4
160	$R3m$	$R3m$	C_{3v}^5
161	$R3c$	$R3c$	C_{3v}^6
162	$P\bar{3}1m$	$P\bar{3}12/m$	D_{3d}^1
163	$P\bar{3}1c$	$P\bar{3}12/c$	D_{3d}^2
164	$P\bar{3}m1$	$P\bar{3}2/m1$	D_{3d}^3
165	$P\bar{3}c1$	$P\bar{3}2/c1$	D_{3d}^4
166	$R\bar{3}m$	$R\bar{3}2/m$	D_{3d}^5
167	$R\bar{3}c$	$R\bar{3}2/c$	D_{3d}^6

Hexagonal

	H.M. (short)	H.M. (full)	S
168	$P6$	$P6$	C_6^1
169	$P6_1$	$P6_1$	C_6^2
170	$P6_5$	$P6_5$	C_6^3
171	$P6_2$	$P6_2$	C_6^4
172	$P6_4$	$P6_4$	C_6^5
173	$P6_3$	$P6_3$	C_6^6
174	$P\bar{6}$	$P\bar{6}$	C_{3h}^1
175	$P6/m$	$P6/m$	C_{6h}^1
176	$P6_3/m$	$P6_3/m$	C_{6h}^2
177	$P622$	$P622$	D_6^1
178	$P6_122$	$P6_122$	D_6^2
179	$P6_522$	$P6_522$	D_6^3
180	$P6_222$	$P6_222$	D_6^4
181	$P6_422$	$P6_422$	D_6^5
182	$P6_322$	$P6_322$	D_6^6
183	$P6mm$	$P6mm$	C_{6v}^1
184	$P6cc$	$P6cc$	C_{6v}^2
185	$P6_3cm$	$P6_3cm$	C_{6v}^3
186	$P6_3mc$	$P6_3mc$	C_{6v}^4
187	$P\bar{6}m2$	$P\bar{6}m2$	D_{3h}^1
188	$P\bar{6}c2$	$P\bar{6}c2$	D_{3h}^2
189	$P\bar{6}2m$	$P\bar{6}2m$	D_{3h}^3
190	$P\bar{6}2c$	$P\bar{6}2c$	D_{3h}^4
191	$P6/mmm$	$P6/m2/m2/m$	D_{6h}^1
192	$P6/mcc$	$P6/m2/c2/c$	D_{6h}^2
193	$P6_3/mcm$	$P6_3/m2/c2/m$	D_{6h}^3
194	$P6_3mmc$	$P6_3/m2/m2/c$	D_{6h}^4

Space Group No.	H.M. (short)	H.M. (full)	S
Cubic			
195	P23	P23	T^1
196	F23	F23	T^2
197	I23	I23	T^3
198	$P2_1\,3$	$P2_1\,3$	T^4
199	$I2_1\,3$	$I2_1\,3$	T^5
200	$Pm\bar{3}$	$P2/m\bar{3}$	T_h^1
201	$Pn\bar{3}$	$P2/n3$	T_h^2
202	$Fm\bar{3}$	$F2/m3$	T_h^3
203	$Fd\bar{3}$	$F2/d3$	T_h^4
204	$Im\bar{3}$	$I2/m3$	T_h^5
205	$Pa\bar{3}$	$P2_1/a3$	T_h^6
206	Ia3	$I2_1/a3$	T_h^7
207	P432	P432	O^1
208	$P4_2\,32$	$P4_2\,32$	O^2
209	F432	F432	O^3
210	$F4_1\,32$	$F4_1\,32$	O^4
211	I432	I432	O^5
212	$P4_3\,32$	$P4_3\,32$	O^6
213	$P4_1\,32$	$P4_1\,32$	O^7
214	$I4_1\,32$	$I4_1\,32$	O^8
215	$P\bar{4}3m$	$P\bar{4}3m$	T_d^1
216	$F\bar{4}3m$	$F\bar{4}3m$	T_d^2
217	$I\bar{4}3m$	$I\bar{4}3m$	T_d^3
218	$P\bar{4}3n$	$P\bar{4}3n$	T_d^4
219	$F\bar{4}3c$	$F\bar{4}3c$	T_d^5
220	$I\bar{4}3d$	$I\bar{4}3d$	T_d^6
221	Pm3m	$P4/m\bar{3}2/m$	O_h^1
222	Pn3n	$P4/n\bar{3}2/n$	O_h^2
223	Pm3n	$P4_2/m\bar{3}2/n$	O_h^3
224	Pn3m	$P4_2/n\bar{3}2/m$	O_h^4
225	Fm3m	$F4/m\bar{3}2/m$	O_h^5
226	Fm3c	$F4/m\bar{3}2/c$	O_h^6
227	Fd3m	$F4_1/d\bar{3}2/m$	O_h^7
228	Fd3c	$F4_1/d\bar{3}2/c$	O_h^8
229	Im3m	$I4/m\bar{3}2/m$	O_h^9
230	Ia3d	$I4_1/a\bar{3}2/d$	O_h^{10}

The directions along which we must list the symmetry elements vary with the crystal class as described below.

Triclinic – only one symbol is required because it represents all directions in space. As we have seen previously the symbol must be either 1 or $\bar{1}$ depending on whether there is or is not a centre of symmetry in the system. P1 and P$\bar{1}$ therefore represent the two triclinic space groups.

Monoclinic – only the one symbol is required and that describes the nature of the unique axis. The following notations represent typical monoclinic space groups:

P2   — a primitive cell in which the unique axis is a 2-fold rotation axis

$P2_1$ — a primitive cell in which the unique axis is a 2_1 screw axis

Cm   — a centred cell in which the unique axis is a 2-fold rotor-inversion axis

Pc   — a primitive cell in which there is a glide plane perpendicular to the unique b axis with a glide of $c/2$

C2/m— a centred cell in which the unique 2-fold axis has a mirror plane normal to it.

In a monoclinic lattice with b as the unique axis the symmetry about both a and c is 1. The space group symbols could be written in full to give the symmetry about all three axes, e.g. P2 could be written as P121. This is not normally done because the 1-fold axes along a and c are redundant symmetry elements. It may, however, be helpful to remember that we are in effect specifying the symmetry about all three axes in writing down a monoclinic space group symbol.

Orthorhombic — in the orthorhombic system the lattice type symbol is followed by three other symbols which represent the symmetries about all three of the orthorhombic axes in the order a,b,c. Typical orthorhombic space groups are:

$P2_1 2_1 2_1$ — primitive cell with screw axes along a, b and c

C222   — cell centred on the C-face with 2-fold rotation axes along a, b and c

Iba2   — body centred cell in which the symmetries about the three axes are:
(a) axial glide plane perpendicular to a with a glide of $b/2$
(b) axial glide plane perpendicular to b with a glide of $a/2$
(c) the c axis is a 2-fold rotation axis

Fdd2   — a cell centred on all faces in which the symmetries about the three axes are:
(a) diamond glide plane perpendicular to a with a glide of $\dfrac{b+c}{4}$

(b) diamond glide plane perpendicular to b with a glide of $\dfrac{a+c}{4}$

(c) the c axis is a 2-fold rotation axis

Pnma — a primitive cell in which the symmetries about the three axes are:

(a) diagonal glide plane perpendicular to a with a glide of $\dfrac{b+c}{2}$

(b) a mirror plane normal to b

(c) axial glide plane perpendicular to c with a glide of $a/2$

Tetragonal – The lattice type symbol is followed by three symbols which represent respectively:

(i) the symmetry of the 4-fold axis, i.e. the c axis

(ii) the symmetry about the a and b axes. The a and b axes must have the same symmetry in the tetragonal system

(iii) the symmetry along the 110 and $\overline{1}\overline{1}0$ planes (i.e. the square diagonals).

Typical tetragonal space groups are:

P4nc — a primitive cell in which the c axis is a 4-fold rotation axis in which there are diagonal glide planes perpendicular to the a and b axes with glides of $\dfrac{b+c}{2}$ and $\dfrac{a+c}{2}$ respectively and in which there is an axial glide plane perpendicular to the square diagonals with a glide of $c/2$.

I4/mmm — a body-centred cell in which the c axis is a 4-fold rotation axis with a mirror plane normal to it and in which there are mirror planes normal to the a and b axes and the square diagonals.

Hexagonal – Hexagonal lattices can be described in terms of a four co-ordinate system x,y,z,u where x,y and z are the a,b and c axes respectively and u lies in the ab plane as shown in Figure 4.10. In the hexagonal system the lattice type symbol is followed by three other symbols which represent:

(i) the symmetry of the 6-fold axis, the c axis, or, if the lattice is rhombohedral the symmetry of the 3-fold axis

(ii) The symmetry about the a,b and u axes

(iii) the symbols for the diad axes normal to the a,b and u axes in the 0001 plane.

A typical hexagonal space group is C6mm and a typical rhombohedral space group is P312.

Cubic – In cubic symmetry the three symbols following the lattice type symbol represent the symmetries in:

(i) the 100 planes, i.e. the three equal cell translations

(ii) the 111 planes, i.e. the cube diagonals

(iii) the 110 planes, i.e. the square diagonals

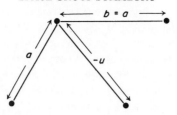

Figure 4.10   The a, b and u axes of a hexagonal lattice in the plane perpendicular to the c axis

Typical cubic space groups are Fm3m, I432 and $P4_3 32$.

In some space groups it is not necessary to list all of the symbols. This can arise when the symbol has only the minimum symmetry about the last direction or directions listed. For example the tetragonal space group P4 is completely described by these symbols and there is no need to specify the symmetries about the 100 and 110 planes. Table 4.8 is a list of the 230 space groups and it contains the Hermann-Mauguin (HM) symbols, the symbols for the full symmetry elements and the Schoenflies (S) notation.

It is often convenient to consider the plane symmetry obtained when a three-dimensional array is projected on to one of its faces. This two-dimensional symmetry gives rise to the 17 plane groups listed in Table 4.9. These plane groups arise from the application of the point groups listed to the plane lattices

Table 4.9. The 17 plane groups

No.	Lattice		Point Groups	Plane Groups
1	Oblique	(p)	1	p1
2		(p)	2	p2
3	Rectangular	(p)	m	pm
4		(p)	m	pg
5		(c)	m	cm
6		(p)	2mm	pmm
7		(p)	2mm	pmg
8		(p)	2mm	pgg
9		(c)	2mm	cmm
10	Square	(p)	4	p4
11		(p)	4mm	p4m
12		(p)	4mm	p4g
13	Hexagonal	(p)	3	p3
14		(p)	3m	p3m1
15		(p)	3m	p31m
16		(p)	6	p6
17		(p)	6mm	p6m

of Figure 4.3. The plane group notations are similar to the space group notations in that they have first the symbol for the plane lattice type (p or c depending upon the type of plane lattice — primitive or centred). This symbol is then followed by symbols representing the symmetries in the plane of the projection (the ab plane). In two-dimensional space all glide planes are denoted by the symbol g. This is because the translation of the glide is fixed by the position of the glide plane. For example, in two dimensions a glide plane perpendicular to a must have a glide of $b/2$ because b is the only other translation.

Space Group Diagrams and Equivalent Positions Diagrams

SPACE GROUP DIAGRAMS

Space group diagrams, like stereographic projections, are designed to present all of the symmetry elements of a space group in a diagrammatic form. The diagrams are usually ab projections bounded by a thin line unless the a or b directions contain mirror or glide planes. The diagrams have the b axis horizontal with its positive direction from right to left. The positive direction of the a axis is from the top to the bottom of the diagram. The symbols used in these diagrams are given in Table 4.10.

Table 4.10. Symbols for space group diagrams

Symmetry element	Diagrammatic symbols for element	
	Normal to the plane of projection	Parallel to the plane of projection
m		
a, b		
c		none
n		
rotation axes	closed polygons as before	
rotor-reflection axes	open polygons as before	
screw axes	polygons with extended sides as before	
rotor-inversion axes	sometimes closed polygons with a small open circle at the centre are used, e.g. ◊ ▲ etc.	

The space group diagrams for P1, P$\bar{1}$, P2$_1$2$_1$2, Pmm2, Pna2$_1$, Pmmm, P4$_2$/n and P4mm are shown in Figure 4.11 and explained below:

    (a) P1   There are no symmetry elements in this space group other than the identity and so the space group diagram is simply an oblique polygon with its sides drawn in as thin lines.

(b) P$\bar{1}$ The centre of symmetry in this group is shown by the presence of the 2-fold rotor-reflection axes in the correct positions within an oblique polygon. The sides of this polygon are again drawn in as thin lines.

(c) P$2_1 2_1 2$ The space group diagram for this symmetry is a rectangular polygon again with its sides drawn in with thin lines. The 2-fold axis along c (normal to the plane of the projection) is shown by the solid polygon symbols for the 2-fold axes. The positions of the 2_1 screw axes along the a and b directions are shown by the half arrows.

(d) Pmm2 The space group diagram shows the two mirror planes parallel to the a and b directions meeting at a 2-fold axis in the centre. The positions of the related 2-fold axes and mirror planes (along the sides of the rectangular diagram) are also shown.

(e) Pna2_1 shows the use of symbols for n (normal to the a direction), a (normal to the b direction) and 2_1 along the c axis.

(f) Pmmm This diagram is similar to that for Pmm2 except that the additional mirror plane parallel to the ab plane is shown by the rectangular symbol at the top right-hand corner. The 2-fold axes along the a and b directions show where pairs of mirror planes meet and are indicated by arrows.

(g) P4_2/n The positions of the 4_2 axes along c and of implied 4-fold rotor-reflection axes are shown. This space group is centrosymmetric and the centres of symmetry are at the positions shown by the 2-fold rotor-reflection symbols. The 1/4 beside this symbol indicates that these centres are at a height of $c/4$ above the ab plane. The symbol for the diagonal glide plane parallel to the ab plane is shown in the top right-hand corner of the diagram as this is also at a height of $c/4$.

(h) P4mm The positions of the 4-fold axes along c, and the mirror planes normal to the a and b directions and to the square diagonals are shown. The glide plane and the 2-fold rotation axes are symmetries which result from the 4mm.

EQUIVALENT POSITIONS DIAGRAMS

As in the case of the stereographic projections of the point groups, we can produce equivalent positions diagrams for the space groups. These diagrams are explained for the eight space groups of Figure 4.11 and are illustrated in Figure 4.12. The general position x,y,z is shown on the diagram as an open circle o followed by + or − to distinguish between positions which are +z (i.e. above the plane of the projection) or −z (below the plane of projection). The open circle is also assumed to be right-handed, if the position is inverted through a centre it would become left-handed and would be shown as an open circle containing a comma, ⊙. The symmetry elements are not included in the equivalent positions

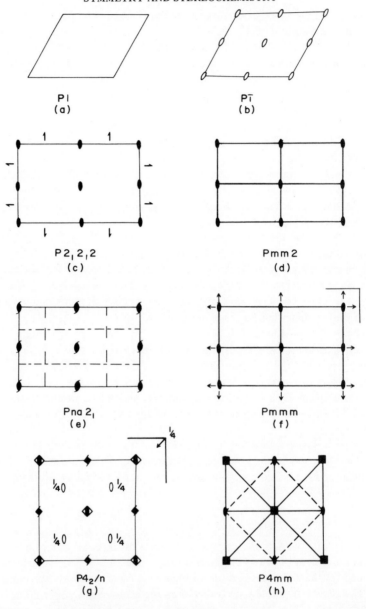

Figure 4.11   Space group diagrams for (a) P1, (b) P$\bar{1}$, (c) P2$_1$2$_1$2, (d) Pmm2, (e) Pna2$_1$
(f) Pmmm, (g) P4$_2$/n (h) P4mm.
The b axis is horizontal with its positive direction from left to right. The a
axis lies up and down the page with its positive direction from the top to
the bottom of the diagram

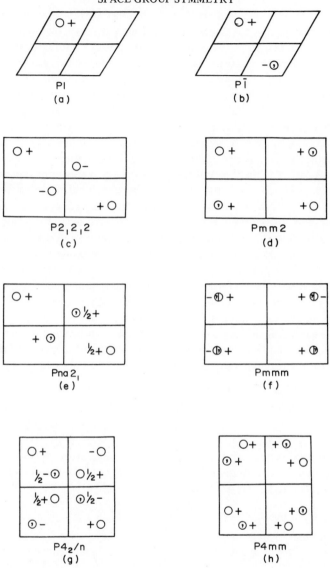

Figure 4.12   Equivalent positions diagrams for the space groups (a) Pl, (b) P$\bar{1}$, (c) P2$_1$2$_1$2, (d) Pmm2, (e) Pna2$_1$, (f) Pmmm, (g) P4$_2$/n (h) P4mm
The axes are defined in the same way as in Figure 4.11

diagrams which are generally divided into four quadrants for convenience.

The equivalent positions diagram for P1 (Figure 3.12a) shows that if we take the general position x,y,z, there is no equivalent position within the cell. This means that if we place an atom centre at x,y,z, there is no other atom in the unit cell that is related to it by symmetry. For P1, therefore, there is only one general position x,y,z. If we place an atom in the general position x,y,z in $P\bar{1}$ it must be inverted by the centre of symmetry and also appear at $-x,-y,-z$. There are thus two equivalent positions in the space group $P\bar{1}$ and these are x,y,z and $-x,-y,-z$. This means that, if we place an atom at the position x,y,z, there must be an identical atom in the position $-x,-y,-z$. It is interesting to consider the effect of placing an atom at one of the centres of symmetry, for example at 0,0,0. Since this atom lies on the centre of symmetry, it has no related equivalent position (identical atom within the cell) and is therefore a *Special Position* of the space group. Space group diagrams always show the general equivalent positions, that is the maximum number of sites within the cell that are symmetry related. It is generally a simple and worthwhile task, however, to find out the effect of placing an atom on one of the symmetry elements of the group. Volume I of the International Tables for X-ray Crystallography (published for the International Union of Crystallography by the Kynoch Press, Birmingham, England) contains full details of all space groups with the relevant diagrams and lists of the general equivalent positions and all special equivalent positions. In this chapter we shall list the general equivalent positions and some of the special positions for the remaining space groups in Figure 4.12 (i.e. 4.12 c to 4.12 h).

$P2_1 2_1 2$ If we place an atom in the general position x,y,z, the symmetry elements of this group generate four equivalent positions at:

$$x,y,z; \quad -x, -y,z; \quad (\tfrac{1}{2} + x), (\tfrac{1}{2} - y), -z; \quad \text{and } (\tfrac{1}{2} - x), (\tfrac{1}{2} + y), -z.$$

These equivalent positions are shown in Figure 4.12(c). If we place an atom on the 2-fold axis it can have any value of z, but the co-ordinates of x and y are limited to either 0, 0 or $0,\tfrac{1}{2}$. There are thus two special 2-fold equivalent positions for this group at

(i) $0,\tfrac{1}{2},z$ and $\tfrac{1}{2},0,-z$

(ii) $0,0,z$ and $\tfrac{1}{2},\tfrac{1}{2},-z$

That is, if we place an atom on the 2-fold axis there can only be one other atom in the cell which is related to it by the symmetry of the group. This can be seen either by drawing an equivalent positions diagram with an atom at 0, 0, z or simply by substituting the values $x = 0$, $y = 0$, $z = z$ in the list of general equivalent positions which will then reduce to only two different positions.

Pmm2 This space group has a 4-fold set of general positions (Figure 4.12 d) at

$$x,y,z; \quad -x,-y,z; \quad x,-y,z; \quad -x,y,z$$

There are 2-fold special positions if an atom lies on one of the mirror planes (e.g. $x,0,z$) and 1-fold special positions if the atom lies at the intersection of two of the mirror planes (e.g. $0,0,z$).

$Pna2_1$ The equivalent general positions (Figure 4.12e) are

$$x,y,z; \quad -x,-y,(\tfrac{1}{2}+z); \quad (\tfrac{1}{2}-x),(\tfrac{1}{2}+y),(\tfrac{1}{2}+z); \quad (\tfrac{1}{2}+x),(\tfrac{1}{2}-y),z$$

There are no special positions. The $\tfrac{1}{2}$ + symbol in the diagram indicates that the co-ordinates of the points in the c direction are $\tfrac{1}{2}+z$.

$Pmmm$ The eight equivalent positions of this group are

$$x,y,z; \quad -x,-y,z; \quad x,-y,-z; \quad -x,y,-z; \quad -x,-y,-z; \quad x,y,-z; \quad -x,y,z; \quad x,-y,z.$$

The symbols in Figure 4.12(f) show that there are two equivalent positions with the same x and y co-ordinates but differing z co-ordinates. For the above symbol this means that there is a right-handed object at x,y,z and a left-handed object at $x,y,-z$.

$P4_2/n$ The eight equivalent general positions of this tetragonal space group are:

$$x,y,z; \quad -x,-y,z; \quad (\tfrac{1}{2}+x),(\tfrac{1}{2}+y),(\tfrac{1}{2}-z); \quad (\tfrac{1}{2}-x),(\tfrac{1}{2}-y),(\tfrac{1}{2}-z); \quad -y,x,-z;$$
$$y,-x,-z; \quad (\tfrac{1}{2}-y),(\tfrac{1}{2}+x),(\tfrac{1}{2}+z); \quad (\tfrac{1}{2}+y),(\tfrac{1}{2}-x),(\tfrac{1}{2}+z).$$

$P4mm$ The eight equivalent general positions of this tetragonal space group are:

$$x,y,z; \quad -x,-y,z; \quad -x,y,z; \quad x,-y,z; \quad y,x,z; \quad -y,-x,z; \quad -y,x,z; \quad y,-x,z$$

The importance of a knowledge of space group symmetry (and of the equivalent positions generated by the symmetry elements present) in the determination of crystal structures will be discussed in Chapter 6.

PROBLEMS

1 With the aid of suitable diagrams show that a C-centred tetragonal cell is equivalent to a primitive tetragonal cell.

2 To which crystal class do each of the following space groups belong. Explain the meanings of the symbols in each case.

(a) $Pcc2$, (b) $P4cc$, (c) $P3$, (d) $Ama2$, (e) $F432$, (f) $P2_1/m$, (g) $P6_322$, (h) $I4_1cd$.

3 Draw the plane group diagrams for the following two-dimensional plane groups:

(i) P1 (oblique)   (ii) Pm (rectangular)
(iii) Pg (rectangular)   (iv) Pmm (rectangular)
(v) P2 (oblique)   (vi) P4 (square)

4 Draw the equivalent positions diagrams for the plane groups listed in question 3 above

5 What are the co-ordinates of equivalent atoms for the special positions which result from placing an atom in the positions:
  (i)   on one of the mirror planes in Pmmm (Figures 4.11 f and 4.12 f) say on (a) $x,0,z$ or (b) $x,y,0$.
  (ii) on the intersection of two of the mirror planes in Pmmm, say on (a) $\frac{1}{2},\frac{1}{2},z$ or (b) $\frac{1}{2},y,0$.
  (iii) on the intersection of three of the mirror planes in Pmmm say on 0,0,0.
  (iv) at the centre of symmetry in $P4_2/n$ (Figures 4.11 g and 4.12 g) say at 1/4,1/4,1/4.
  (v)   on a $\bar{4}$ axis in $P4_2/n$ say on 0,0,0.
  (vi) on a mirror plane in P4mm (Figures 4.11 h and 4.12 h) say on $x,0,z$.
  (vii) at the junction of the 4-fold axis and the two mirror planes of P4mm say on $0,0,z$.

6 Explain why the symbols P321 and P312 represent different space groups while $P2_1 22$ and $P222_1$ represent the same symmetry.

7 Which of the following pairs of space group symbols represent different space groups:
  (a) P2   , C2
  (b) P$\bar{6}$m2   , P$\bar{6}$2m
  (c) Pcca   , Pccb
  (d) P$4_1$32 , P432
  (e) C222   , A222

5

Group Theory

Symmetry Operations as a Group

In Chapter 2 we saw the effect of subjecting a molecule to successive symmetry operations, and collected the information into a multiplication table, of which Tables 2.1, 2.2 and 3.2 are examples. If we study these tables, we see that the following rules apply.

(i) Whenever we perform two successive symmetry operations, A, B on a molecule, the result is the same as would have been produced by some single symmetry operation of the molecule. This is the so-called *Combinative Law*.

(ii) If we perform three symmetry operations, A, B, C, we get the same result whether we perform C followed by the product AB, or the product BC followed by A.

This is the so-called *Associative Law* $(AB)C = A(BC)$

(iii) For every operation A there is an inverse which we can designate A^{-1}, such that $A^{-1}A = AA^{-1} = I$, the identity.

(iv) The operation I exists, and for every operation A, we have $AI = IA = A$.

Since the symmetry operations of a molecule obey the four rules given above, they form a *Group*. Note that, although the designation 'point group' comes from the fact that all the symmetry elements of a molecule pass through a point, the group consists of the operations, not the elements. Confusion often arises here because the members of a group (i.e. the symmetry operations, in this case) are sometimes referred to as the elements of the group.

As we stated in Chapter 2 the multiplication tables for the symmetry operations of a molecule contain the information we need in order to simplify physical problems by considering them in terms of molecular symmetry.

To transform this information into a more readily usable form, we can try to find a set of functions which multiply in the same way as the symmetry operations. Suppose we rewrite Table 2.1 as Table 5.1 replacing I, C_2, σ_h, i by A, B, C, D, can we find four functions A, B, C, D such that the multiplication table is still true?

Table 5.1. Another version of the multiplication table of the point group C_{2h}, previously given as Table 2.1

	A	B	C	D
A	A	B	C	D
B	B	A	D	C
C	C	D	A	B
D	D	C	B	A

If we can find such a set of functions, that set is said to represent the symmetry operations. The four rules given above, combined with the fact that, in general, multiplication of the symmetry operations is not commutative, suggest that a matrix would be a suitable type of function to represent a symmetry operation.

If we are to use matrices to represent symmetry operations, we need to know how to multiply matrices. If we have two matrices $O, R,$ their product P is formed by multiplying, term by term, each row of Q into the corresponding column of R, as illustrated in the example below. This requirement means that it is not always possible to form a matrix product; matrices which can be multiplied together are said to be conformable. A matrix is simply an array of numbers written as rows and columns, thus

$$\begin{bmatrix} 2 & 1 & 3 \\ 5 & 7 & 6 \end{bmatrix}$$

is a matrix and is said to be rectangular since it has two rows and three columns, while

$$\begin{bmatrix} 4 & 1 \\ 3 & 7 \end{bmatrix}$$

is said to be square since it has two rows and two columns.

All matrices which can represent symmetry operations are square.

MATRIX MULTIPLICATION

The conformability condition is usually stated as follows:

For matrices to be conformable, the number of rows in one must be the same as the number of columns in the other. To be more specific, one should say that the product QR can be formed when the number of columns in Q is equal to the number of rows in R, while the product RQ can be formed when the number of columns in R is equal to the number of rows in Q.

So for:

(i) $Q = \begin{bmatrix} 2 \\ 1 \end{bmatrix}$, $R = \begin{bmatrix} 3 & 4 \\ 6 & 5 \end{bmatrix}$; $\begin{bmatrix} q_{11} \\ q_{21} \end{bmatrix} \begin{bmatrix} r_{11} & r_{12} \\ r_{21} & r_{22} \end{bmatrix}$

We can form RQ but not QR.

(ii) $Q = \begin{bmatrix} 2 & 3 \\ 1 & 6 \end{bmatrix}$ $R = \begin{bmatrix} 4 \\ 5 \end{bmatrix}$; $\begin{bmatrix} q_{11} & q_{12} \\ q_{21} & q_{22} \end{bmatrix} \begin{bmatrix} r_{11} \\ r_{21} \end{bmatrix}$

We can form QR but not RQ.

(iii) $Q = \begin{bmatrix} 3 & 4 \\ 5 & 6 \end{bmatrix}$ $R = \begin{bmatrix} 1 & 2 \\ 7 & 8 \end{bmatrix}$.

We can form both QR and RQ.

(a) for (i) $P = RQ \therefore P_{ij} = \sum_k r_{ik} q_{kj}$

$$= \begin{bmatrix} r_{11}q_{11} + r_{12}q_{21} \\ r_{21}q_{11} + r_{22}q_{21} \end{bmatrix} = \begin{bmatrix} 3 \times 2 + 4 \times 1 \\ 6 \times 2 + 5 \times 1 \end{bmatrix} = \begin{bmatrix} 10 \\ 17 \end{bmatrix}$$

(b) for (ii) $P = QR \therefore p_{ij} = \sum_k q_{ik} r_{kj}$

$$= \begin{bmatrix} q_{11}r_{11} + q_{12}r_{21} \\ q_{21}r_{11} + q_{22}r_{21} \end{bmatrix} = \begin{bmatrix} 2 \times 4 + 3 \times 5 \\ 1 \times 4 + 6 \times 5 \end{bmatrix} = \begin{bmatrix} 23 \\ 34 \end{bmatrix}$$

In the co-ordinate transformation due to clockwise rotation by an angle θ, the transformation matrix is correctly expressed as

$$\underset{P}{\begin{bmatrix} x' \\ y' \end{bmatrix}} = \underset{Q}{\begin{bmatrix} \cos\theta & \sin\theta \\ -\sin\theta & \cos\theta \end{bmatrix}} \underset{R}{\begin{bmatrix} x \\ y \end{bmatrix}}$$

This is illustrated by Figure 5.1

where $x' = x\cos\theta + y\sin\theta$
$y' = -x\sin\theta + y\cos\theta$
compare this with example (ii) above.

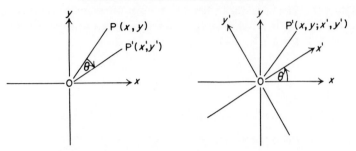

Figure 5.1(a)   Rotation of OP through θ in a clockwise direction is equivalent to rotation of the axes through θ in an anti-clockwise direction,

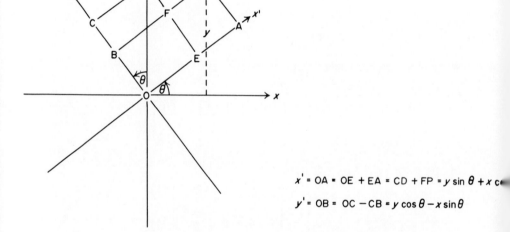

$$x' = OA = OE + EA = CD + FP = y \sin \theta + x \cos\theta$$

$$y' = OB = OC - CB = y \cos \theta - x \sin\theta$$

Figure 5.1(b)   On rotation of the axes through θ in an anti-clockwise direction, the new co-ordinates of P are $x' = x \cos\theta + y \sin\theta$, $y' = -x \sin\theta + y \cos\theta$

By the rule we have $p_{ij} = \sum_k q_{ik} r_{kj}$

$$\therefore \quad p_{11} = q_{11} r_{11} + q_{12} r_{21} = \cos\theta.x + \sin\theta.y$$
$$p_{21} = q_{21} r_{21} + q_{22} r_{21} = -\sin\theta.x + \cos\theta.y$$

Also note that, if one matrix has m rows and n columns, while the other has n rows and p columns, their product will have m rows and p columns. In the co-ordinate transformation matrix, Q has 2 rows (m) and 2 columns (n), while R has 2 rows (n) and 1 column (p), so P has 2 rows (m) and 1 column (p).

MATRICES AS REPRESENTATIONS OF SYMMETRY OPERATIONS

We may illustrate the representation of symmetry operations by matrices using Table 2.2, the multiplication table for the NH_3 molecule.

We see that
$$\sigma_{v1}\sigma_{v2} = C_3^1$$
$$\sigma_{v2}\sigma_{v1} = C_3^2$$

Now since the matrix expressing the effect of a rotation through an angle θ is

$$\begin{bmatrix} \cos\theta & \sin\theta \\ -\sin\theta & \cos\theta \end{bmatrix}$$

the matrix representing a rotation through $120°$, that is, the operation C_3^1, is

$$\begin{bmatrix} \cos 120° & \sin 120° \\ -\sin 120° & \cos 120° \end{bmatrix} = \begin{bmatrix} -1/2 & \sqrt{3}/2 \\ -\sqrt{3}/2 & -1/2 \end{bmatrix}$$

while that representing a rotation through $240°$ (C_3^2) is

$$\begin{bmatrix} \cos 240° & \sin 240° \\ -\sin 240° & \cos 240° \end{bmatrix} = \begin{bmatrix} -1/2 & -\sqrt{3}/2 \\ \sqrt{3}/2 & -1/2 \end{bmatrix}$$

Now consider the matrices

$$Q = \begin{bmatrix} 1 & 0 \\ 0 & -1 \end{bmatrix} \quad R = \begin{bmatrix} -1/2 & \sqrt{3}/2 \\ \sqrt{3}/2 & 1/2 \end{bmatrix}$$

The product QR is

$$\begin{bmatrix} -1/2 & \sqrt{3}/2 \\ -\sqrt{3}/2 & -1/2 \end{bmatrix}$$

and the product RQ is

$$\begin{bmatrix} -1/2 & -\sqrt{3}/2 \\ \sqrt{3}/2 & -1/2 \end{bmatrix}$$

Since QR is the matrix expressing C_3^1, and RQ is that expressing C_3^2, Q and R can be used to represent the symmetry operations σ_{v1} and σ_{v2}.

In practice the problems we have to solve do not depend on determining the actual matrices which make up a representation of the symmetry operations; we need to know only the characters of the matrices. The character (χ) is the sum of the diagonal elements of the matrix, and is also referred to as the trace or spur.

The characters of the matrices Q, R, QR, and RQ are $0, 0, -1$ and -1 respectively. In any case, there is an infinite number of representations of a group of symmetry operations, but only a very few of these are of interest. One such representation is formed by considering the displacements of the atoms of a molecule, in Cartesian co-ordinates. Figure 5.2 illustrates this for a triatomic molecule AB_2 of C_{2v} symmetry. The symmetry operations are I, C_2, $\sigma_v(yz)$ and $\sigma_v(xz)$. These two reflections are the reflections at the molecular plane and at the mirror plane perpendicular to the molecular plane, respectively. We may express the effects of the symmetry operations by drawing up a transformation table.

Table 5.2. Transformation table for the Cartesian displacements of AB_2 with C_{2v} symmetry

Operation	I	C_2	$\sigma_v(yz)$	$\sigma_v(xz) = \sigma_v'$
Co-ordinate				
x_1	x_1	$-x_1$	$-x_1$	x_1
y_1	y_1	$-y_1$	y_1	$-y_1$
z_1	z_1	z_1	z_1	z_1
x_2	x_2	$-x_3$	$-x_2$	x_3
y_2	y_2	$-y_3$	y_2	$-y_3$
z_2	z_2	z_3	z_2	z_3
x_3	x_3	$-x_2$	$-x_3$	x_2
y_3	y_3	$-y_2$	y_3	$-y_2$
z_3	z_3	z_2	z_3	z_2

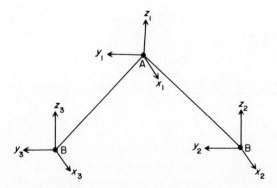

Figure 5.2   Cartesian displacements for the atoms of a molecule AB_2 with C_{2v} symmetry. The C_2 axis lies in the direction of z_1 and the plane of the molecule is the yz plane

For each operation, the relation between the original co-ordinate and the transformed co-ordinate (designated by a prime) produced by the operation can be expressed in matrix form, as follows:

$$A = I$$

$$
\begin{bmatrix} x_1' \\ y_1' \\ z_1' \\ x_2' \\ y_2' \\ z_2' \\ x_3' \\ y_3' \\ z_3' \end{bmatrix}
=
\begin{bmatrix}
1 & 0 & 0 & 0 & 0 & 0 & 0 & 0 & 0 \\
0 & 1 & 0 & 0 & 0 & 0 & 0 & 0 & 0 \\
0 & 0 & 1 & 0 & 0 & 0 & 0 & 0 & 0 \\
0 & 0 & 0 & 1 & 0 & 0 & 0 & 0 & 0 \\
0 & 0 & 0 & 0 & 1 & 0 & 0 & 0 & 0 \\
0 & 0 & 0 & 0 & 0 & 1 & 0 & 0 & 0 \\
0 & 0 & 0 & 0 & 0 & 0 & 1 & 0 & 0 \\
0 & 0 & 0 & 0 & 0 & 0 & 0 & 1 & 0 \\
0 & 0 & 0 & 0 & 0 & 0 & 0 & 0 & 1
\end{bmatrix}
\begin{bmatrix} x_1 \\ y_1 \\ z_1 \\ x_2 \\ y_2 \\ z_2 \\ x_3 \\ y_3 \\ z_3 \end{bmatrix}
$$

$$B = C_2$$

$$
\begin{bmatrix} x_1' \\ y_1' \\ z_1' \\ x_2' \\ y_2' \\ z_2' \\ x_3' \\ y_3' \\ z_3' \end{bmatrix}
=
\begin{bmatrix}
1 & 0 & 0 & 0 & 0 & 0 & 0 & 0 & 0 \\
0 & -1 & 0 & 0 & 0 & 0 & 0 & 0 & 0 \\
0 & 0 & 1 & 0 & 0 & 0 & 0 & 0 & 0 \\
0 & 0 & 0 & 0 & 0 & 0 & -1 & 0 & 0 \\
0 & 0 & 0 & 0 & 0 & 0 & 0 & -1 & 0 \\
0 & 0 & 0 & 0 & 0 & 0 & 0 & 0 & 1 \\
0 & 0 & 0 & -1 & 0 & 0 & 0 & 0 & 0 \\
0 & 0 & 0 & 0 & -1 & 0 & 0 & 0 & 0 \\
0 & 0 & 0 & 0 & 0 & 1 & 0 & 0 & 0
\end{bmatrix}
\begin{bmatrix} x_1 \\ y_1 \\ z_1 \\ x_2 \\ y_2 \\ z_2 \\ x_3 \\ y_3 \\ z_3 \end{bmatrix}
$$

SYMMETRY AND STEREOCHEMISTRY

$$C = \sigma_v(yz)$$

$$
\begin{bmatrix} x'_1 \\ y'_1 \\ z'_1 \\ x'_2 \\ y'_2 \\ z'_2 \\ x'_3 \\ y'_3 \\ z'_3 \end{bmatrix}
=
\begin{bmatrix}
-1 & 0 & 0 & 0 & 0 & 0 & 0 & 0 & 0 \\
0 & 1 & 0 & 0 & 0 & 0 & 0 & 0 & 0 \\
0 & 0 & 1 & 0 & 0 & 0 & 0 & 0 & 0 \\
0 & 0 & 0 & -1 & 0 & 0 & 0 & 0 & 0 \\
0 & 0 & 0 & 0 & 1 & 0 & 0 & 0 & 0 \\
0 & 0 & 0 & 0 & 0 & 1 & 0 & 0 & 0 \\
0 & 0 & 0 & 0 & 0 & 0 & -1 & 0 & 0 \\
0 & 0 & 0 & 0 & 0 & 0 & 0 & 1 & 0 \\
0 & 0 & 0 & 0 & 0 & 0 & 0 & 0 & 1
\end{bmatrix}
\begin{bmatrix} x_1 \\ y_1 \\ z_1 \\ x_2 \\ y_2 \\ z_2 \\ x_3 \\ y_3 \\ z_3 \end{bmatrix}
$$

$$D = \sigma_v(xz) = \sigma'_v$$

$$
\begin{bmatrix} x'_1 \\ y'_1 \\ z'_1 \\ x'_2 \\ y'_2 \\ z'_2 \\ x'_3 \\ y'_3 \\ z'_3 \end{bmatrix}
=
\begin{bmatrix}
-1 & 0 & 0 & 0 & 0 & 0 & 0 & 0 & 0 \\
0 & -1 & 0 & 0 & 0 & 0 & 0 & 0 & 0 \\
0 & 0 & 1 & 0 & 0 & 0 & 0 & 0 & 0 \\
0 & 0 & 0 & 0 & 0 & 0 & 1 & 0 & 0 \\
0 & 0 & 0 & 0 & 0 & 0 & 0 & -1 & 0 \\
0 & 0 & 0 & 0 & 0 & 0 & 0 & 0 & 1 \\
0 & 0 & 0 & 1 & 0 & 0 & 0 & 0 & 0 \\
0 & 0 & 0 & 0 & -1 & 0 & 0 & 0 & 0 \\
0 & 0 & 0 & 0 & 0 & 1 & 0 & 0 & 0
\end{bmatrix}
\begin{bmatrix} x_1 \\ y_1 \\ z_1 \\ x_2 \\ y_2 \\ z_2 \\ x_3 \\ y_3 \\ z_3 \end{bmatrix}
$$

Each symmetry operation is represented by a 9 x 9 matrix, and this representation in Cartesian co-ordinates can be written shortly as

	I	C_2	σ_v	σ_v'
Γ Cartesian	9	−1	3	1

where the numbers in each column are the characters of the matrices representing each symmetry operation.

Reducibility and Irreducibility

The representation formed by these four matrices is said to be reducible. This arises as follows:

Let the matrices representing I, C_2, σ_v, σ_v' be labelled A, B, C, D respectively. There is some matrix Q with an inverse Q^{-1} such that

$$Q^{-1}AQ = A'$$
$$Q^{-1}BQ = B'$$
$$Q^{-1}CQ = C'$$
$$Q^{-1}DQ = D'$$

where A', B', C', D' are all blocked out in the same way. In fact, in this case, each of the four matrices A', B', C', D' has non-zero elements only on the leading diagonal. If, after this reduction, no further matrix R (say) can be found such that

$$R^{-1}A'R = A''$$
$$R^{-1}B'R = B''$$
$$R^{-1}C'R = C''$$
$$R^{-1}D'R = D''$$

where A'', B'', C'', D'' are blocked out in the same way then the set of matrices A', B', C', D' forms an irreducible representation. In the case considered here, there is no matrix R leading to a further reduction, so that the matrices A', B', C', D' are irreducible.

As we show later, we do not need to know Q, so only the transformed matrices A', B', C', D' are given and $Q^{-1}AQ$, etc. are not worked out.

The four transformed matrices are

$$A'=(I) \quad \begin{bmatrix} 1 & & & & & & & & \\ & 1 & & & & & & & \\ & & 1 & & & & & & \\ & & & 1 & & & & & \\ & & & & 1 & & & & \\ & & & & & 1 & & & \\ & & & & & & 1 & & \\ & & & & & & & 1 & \\ & & & & & & & & 1 \end{bmatrix} \quad B'=(C_2) \quad \begin{bmatrix} 1 & & & & & & & & \\ & 1 & & & & & & & \\ & & 1 & & & & & & \\ & & & 1 & & & & & \\ & & & & -1 & & & & \\ & & & & & -1 & & & \\ & & & & & & -1 & & \\ & & & & & & & -1 & \\ & & & & & & & & -1 \end{bmatrix}$$

$$C'=(\sigma_v) \quad \begin{bmatrix} 1 & & & & & & & & \\ & 1 & & & & & & & \\ & & 1 & & & & & & \\ & & & -1 & & & & & \\ & & & & -1 & & & & \\ & & & & & -1 & & & \\ & & & & & & 1 & & \\ & & & & & & & 1 & \\ & & & & & & & & 1 \end{bmatrix} \quad D'=(\sigma_v') \quad \begin{bmatrix} 1 & & & & & & & & \\ & 1 & & & & & & & \\ & & 1 & & & & & & \\ & & & -1 & & & & & \\ & & & & 1 & & & & \\ & & & & & 1 & & & \\ & & & & & & -1 & & \\ & & & & & & & -1 & \\ & & & & & & & & -1 \end{bmatrix}$$

If we label the diagonal elements of these four matrices as a'_{jj}, b'_{jj}, c'_{jj}, d'_{jj}, then each set such as a'_{11}, b'_{11}, c'_{11}, d'_{11} is an irreducible representation.

Table 5.3. Irreducible representations formed by the matrices A', B', C', D'

Operation / jj value	I	C_2	σ_v	σ_v'
1, 2 or 3	1	1	1	1
4	1	1	-1	-1
5 or 6	1	-1	-1	1
7, 8 or 9	1	-1	1	-1

Characters of the irreducible representations of the common point groups are given in the character tables collected as Appendix 2, but a few character tables which are required in working out examples are also given in the main text (e.g. Tables 5.4, 5.8, 5.11).

Comparison with the character table of the point group C_{2v} (Table 5.4) shows that all four of the irreducible representations of the group occur; the

Table 5.4. Character table of the point group C_{2v}

	I	C_2	$\sigma_v(yz)$	$\sigma'_v(xz)$
A_1	1	1	1	1
A_2	1	1	−1	−1
B_1	1	−1	−1	1
B_2	1	−1	1	−1

reducible representation formed by the original set of matrices A, B, C, D contains the A_1, irreducible representation three times, the A_2 once, the B_1 twice and the B_2 three times.

Note also that the character of the representation formed by the matrices A', B', C', D' is the same as that formed from the matrices A, B, C, D, that is to say the transformation $A' = Q^{-1}AQ$ leaves the character of A unaltered; similarly for $B' = Q^{-1}BQ$, $C' = Q^{-1}CQ$ and $D' = Q^{-1}DQ$. If we were to find the matrix Q and its inverse Q^{-1}, we would be carrying out an explicit reduction of the set A, B, C, D. We are, however, not normally interested in the nature of Q and Q^{-1}; what we need to know is how many times each irreducible representation of a group occurs in any reducible representation which we construct, and we can do this without finding Q and Q^{-1}

Classes of Operations

We saw earlier when discussing the use of matrices in representing symmetry operations, that the matrices Q and R, which we used to represent reflections at the symmetry planes in the NH_3 molecule, both had characters of 0. Further, the matrices P $(= QR)$ and P' $(= RQ)$ each had the character -1 and represented rotations about the 3-fold axis. In a qualitative way, therefore, it seems that in this case operations of the same type have the same character. We can express this similarity between symmetry operations in a more formal manner by considering what is meant by a class. We know that every element X of a group (i.e. symmetry operation of a molecule) has an inverse X^{-1} such that $XX^{-1} = X^{-1}X = I$, the identity. Now consider Table 5.1, and let each of A, B, C, D be taken as X in turn. Since A represents the identity element I, Table 5.1 shows that every element is its own inverse. Let the inverse of B be called B^{-1}; then $BB^{-1} = B^{-1}B = A$. Table 5.1 shows that if we perform B first, we need to perform B again to get A as the product. Similarly, the operation which has to precede B in order to get A as the product is itself B. The same holds for C and D. We can draw up a table (Table 5.5) showing the result of performing $X^{-1}GX$ where X is A, B, C, D in turn and G is A, B, C, D in turn.

Table 5.5. $X^{-1}GX$ where G, X represent operations of the point group C_{2h}

G⧵X	A	B	C	D
A	A	B	C	D
B	A	B	C	D
C	A	B	C	D
D	A	B	C	D

If X is A and G is A, then $X^{-1}GX = AAA = A$.

If X is A and G is B, then $X^{-1}GX = ABA$.

Now $ABA = A(BA)$ and from Table 5.1, $(BA) = B$ then $A(BA) = AB = B$.

If X is B and G is A, then $X^{-1}GX = BAB = B(AB)$.

From Table 5.1 $(AB) = B$ and $BB = A$.

By repeating this reasoning, the reader may verify the remainder of the results of Table 5.5. The table shows that, for every element G, $X^{-1}GX = G$. Therefore, every element G is said to be a in class by itself. Qualitatively this seems reasonable, since every element is of a different type. A represents I, B represents C_2, C represents σ_h, and D represents i.

Now consider the ammonia molecule, whose multiplication table was given as Table 2.2. If we rewrite Table 2.2 with $I = A$, $C_3^1 = B$, $C_3^2 = C$, $\sigma_{v1} = D$, $\sigma_{v2} = E$, $\sigma_{v3} = F$, the new table (Table 5.6) is as shown.

Table 5.6. Another version of the multiplication table previously given as Table 2.2

	A	B	C	D	E	F
A	A	B	C	D	E	F
B	B	C	A	F	D	E
C	C	A	B	E	F	D
D	D	E	F	A	B	C
E	E	F	D	C	A	B
F	F	D	E	B	C	A

The inverses of the elements are as follows:

$$A^{-1} = A, \; B^{-1} = C, \; C^{-1} = B, \; D^{-1} = D, \; E^{-1} = E, \; F^{-1} = F.$$

We may now construct the table giving $X^{-1}GX$, where X, G are in turn, $A, B, C,$ $D, E, F,$ similarly to Table 5.5.

Table 5.7 $X^{-1}GX$ where X, G represent the symmetry operations of the point group C_{3v}

X \ G	A	B	C	D	E	F
A	A	B	C	D	E	F
B	A	B	C	F	D	E
C	A	B	C	E	F	D
D	A	C	B	D	F	E
E	A	C	B	F	E	D
F	A	C	B	E	D	F

Table 5.7 shows that if $G = A$, $X^{-1}GX$ is A for all X. Therefore A is in a class by itself. If $G =$ either B or C, $X^{-1}GX$ is either B or C. Thus B and C form a class. Finally, if $G = D$, E or F, $X^{-1}GX$ is D, E or F, so that D, E and F form yet another class. Further, the members of any one class are operations of the same type, B and C being C_3^1, C_3^2, while D, E, F are σ_{v1}, σ_{v2}, σ_{v3}. The group contains 6 symmetry operations altogether; the number of symmetry operations is the *Order* of the group (h). There are three classes, and the number of elements in each class is an integral divisor of h. The importance of this division of elements into classes is that all the operations of a class have the same character. Thus we can write the character table of the group C_{3v} in the compact manner of Table 5.8 instead of the extended manner of Table 5.9.

Table 5.8. Character table of the point group C_{3v}

	I	$2C_3$	$3\sigma_v$
A_1	1	1	1
A_2	1	1	−1
E	2	−1	0

Table 5.9. Another form of the character table of the point group C_{3v}

	I	C_3^1	C_3^2	σ_{v1}	σ_{v2}	σ_{v3}
A_1	1	1	1	1	1	1
A_2	1	1	1	−1	−1	−1
E	2	−1	−1	0	0	0

Table 5.4 and 5.8 both illustrate the fact that for any point group, the number of irreducible representations is equal to the number of classes.

LABELLING OF IRREDUCIBLE REPRESENTATIONS

Table 5.4 contains the symbols A_1, A_2, B_1, B_2 which designate the irreducible representations of the point group C_{2v}; Table 5.8 contains A_1, A_2 and E which designate those of C_{3v}. Most chemists use this system, which is due to Mulliken. The meanings of the symbols commonly used are as follows:

A and B each designate a one-dimensional representation, that is, one in which the character of the identity operation is 1. One-dimensional representations which are symmetric to rotation about the rotation axis of highest order are labelled A; those which are antisymmetric to such a rotation are labelled B. Subscripts g, u denote symmetry or antisymmetry with respect to inversion at a centre of symmetry; superscripts $'$, $''$ denote symmetry or antisymmetry with respect to reflection at a horizontal plane, σ_h. For one-dimensional representations only (excluding those of the point groups D_2 and D_{2h}), the subscripts 1 and 2 denote symmetry or antisymmetry with respect to reflection at a vertical (σ_v) or diagonal (σ_d) plane.

For D_2 and D_{2h} we have the symbols B_1, B_2, B_3. These groups have three mutually perpendicular 2-fold axes. Representations labelled A are symmetric with respect to rotation about each of them; those labelled B are symmetric with respect to rotation about one of them only. By convention, B_1, B_2, B_3 denote symmetry with respect to rotation about the z, y and x axes respectively.

For two-dimensional representations denoted by E and for three-dimensional representations denoted by F, the g, u and $'$, $''$ symbols retain the same meaning, but the numerical subscripts do not. It is sufficient for the purpose of the applications dealt with in Chapters 6 and 7 to regard these as arbitrary labels.

Reduction of a Representation

The use of group theory in solving physical problems depends to a large extent on constructing a reducible representation for the property which is of interest, and determining how often each of the irreducible representations of the group occurs in the reducible representation. Although this can be done explicitly, as we have seen, this is normally a very long and complicated process. We can, however, use some of the properties of irreducible representations to derive a simple formula which can be readily applied. The two properties of importance in this respect are

(i) The sum of the squares of the characters in any irreducible representation is equal to the order h of the group

$$\sum_R [\chi_i(R)]^2 = h \qquad (5.1)$$

(ii) We may regard any irreducible representation as a vector with h components, each component being the character of an operation in that representation. Two vectors representing two different irreducible representations (say Γ_i and Γ_j) are orthogonal. If the characters of Γ_i, Γ_j for any operation R are $\chi_i(R)$, $\chi_j(R)$, then, for these vectors to be orthogonal

$$\sum_R \chi_i(R)\chi_j(R) = 0 \text{ when } i \neq j \tag{5.2}$$

Now consider a reducible representation Γ whose character for each operation R is $\chi(R)$. We know that, if the representation is reducible, there is some similarity transformation which will reduce each of the matrices $(A, B, C \ldots)$ making up Γ to another set $(A', B', C' \ldots)$ which is arranged in blocks about the leading diagonal, each corresponding set of blocks making up one irreducible representation of the group. Now $\chi(R)$ is unchanged by a similarity transformation. Let the jth irreducible representation appear a_j times

$$\chi(R) = \sum_j a_j \chi_j(R) \tag{5.3}$$

Multiply both sides of (5.3) by $\chi_i(R)$, where $\chi_i(R)$ is the character of the ith irreducible representation, and sum over all operations R. Then

$$\sum_R \chi(R)\chi_i(R) = \sum_R \sum_j a_j \chi_j(R)\chi_i(R)$$

$$= \sum_j \sum_R a_j \chi_j(R)\chi_i(R) \tag{5.4}$$

For each term in the sum over j, we have

$$\sum_R a_j \chi_j(R)\chi_i(R) = a_j \sum_R \chi_j(R)\chi_i(R)$$

But when $i \neq j$, $\sum_R \chi_j(R)\chi_i(R) = 0$, from (5.2)

and when $i = j$, $\sum_R \chi_j(R)\chi_i(R) = \sum_R [\chi_j(R)]^2 = h$

Thus $\quad \sum_R a_j \chi_j(R)\chi_i(R) = ha_j \delta_{ij}$ where $\delta_{ij} = 1$ when $i = j$, and 0 when $i \neq j$.

(5.5)

Since all terms are zero unless $i = j$, we can rewrite (5.5) as

$$\sum_R \chi(R)\chi_j(R) = ha_j$$

$$\therefore a_j = \frac{1}{h} \sum_R \chi(R)\chi_j(R) \tag{5.6}$$

This equation applies to the character table in its expanded form, where the character for every operation is written down separately. If we use the table in its normal form, we must take account of the fact that an operation will in general occur g_R times, so that (5.6) should be used in the form

$$a_j = \frac{1}{h}\Sigma_R g_R \chi(R)\chi_j(R) \tag{5.7}$$

Thus, to find how many times the jth irreducible representation occurs in a reducible representation, we construct a term of the form

(number of operations in the class times the character of that operation in the reducible representation times the character of the same operation in the jth irreducible representation)

for each operation of the group. The sum of these terms, divided by the order of the group, gives a_j. Let us illustrate this by reducing the representation

I	C_2	$\sigma_v(yz)$	$\sigma_v'(xz)$
9	-1	3	1

of the group C_{2v} which we constructed earlier. The characters of these operations in the four irreducible representations are given in Table 5.4. Since g_R, the number of operations in each class, and $\chi(R)$, the character of each operation in the reducible representation, do not depend on which irreducible representation we are considering, we may begin by writing down $g_R\chi(R)$

	I	C_2	$\sigma_v(yz)$	$\sigma_v'(xz)$
$\chi(R)$	9	-1	3	1
g_R	1	1	1	1
$g_R\chi(R)$	9	-1	3	1

To find how many times the A_1 irreducible representation occurs, we need to multiply each term $g_R\chi(R)$ by $\chi(R)$ for the same operation in the A_1 representation. Table 5.4 shows that for every R, $\chi_{A_1}(R) = 1$, so we have

$$a_{(A_1)} = \frac{1}{h}\Sigma_R g_R\chi(R)\chi_{(A_1)}(R)$$

$$= \frac{1}{4}[(9.1) + (1.-1) + (3.1) + (1.1)] = 3$$

For the A_2 representation, $\chi(I)$ and $\chi(C_2)$ are 1, but $\chi(\sigma_v)$ and $\chi(\sigma_v')$ are -1; thus we have

$$a_{(A_2)} = \frac{1}{4}[(9.1) + (-1.1) + (3.-1) + (1.-1)] = 1$$

Similarly

$$a_{(B_1)} = \frac{1}{4}[(9.1) + (-1.-1) + (3.-1) + (1.1)] = 2$$

$$a_{(B_2)} = \frac{1}{4}[(9.1) + (-1.-1) + (3.1) + (1.-1)] = 3$$

Now let us carry out the reduction of a representation of a group where the classes of symmetry operations contain more than one member. The point group C_{3v} whose character table was given as Table 5.8, is such a group. Let us take the representation Γ whose character $\chi(R)$ is

	I	$2C_3$	$3\sigma_v$
$\chi(R)$	6	0	2

(5.8)

As before, we begin by forming $g_R \chi(R)$

	I	$2C_3$	$3\sigma_v$
$\chi(R)$	6	0	2
g_R	1	2	3
$g_R\chi(R)$	6	0	6

Combining this with the characters of the irreducible representations given in Table 5.8, we have

$$a_{(A_1)} = \frac{1}{6}[(6.1) + (0.1) + (6.1)] = 2$$

$$a_{(A_2)} = \frac{1}{6}[(6.1) + (0.1) + (6.-1)] = 0$$

$$a_{(E)} = \frac{1}{6}[(6.2) + (0.-1) + (6.0)] = 2$$

Therefore, the reducible representation whose character is given by (5.8) above contains the A_1 and E irreducible representations twice each, and does not contain the A_2 irreducible representation. We would write this conclusion as

$$\Gamma = 2A_1 + 2E$$

We can always determine by inspection whether the character of a representation is that of an irreducible representation, because if it is, it will appear in the relevant character table. If it does not appear in the table, there are two possibilities; it is either a reducible representation, or it is not a representation at all. Consider Table 5.1 which symbolises the relations between the symmetry operations of the point group C_{2h}. Suppose we suggest, as a possible representation, $A = 1, B = -1$, $C = 1, D = 1$. Then we can see whether the products formed by these multiplications are those demanded by Table 5.1.

Table 5.10. To show whether $A = 1, B = -1, C = 1, D = 1$ is a representation of C_{2h}

Product	Expected value (Table 5.1)	Actual value
$AA = 1 \times 1$	$A = 1$	$A = 1$
$BA = -1 \times 1$	$B = -1$	$B = -1$
$CA = 1 \times 1$	$C = 1$	$C = 1$
$DA = 1 \times 1$	$D = 1$	$D = 1$
$AB = 1 \times -1$	$B = -1$	$B = -1$
$BB = -1 \times -1$	$A = 1$	$A = 1$
$CB = 1 \times -1$	$D = 1$	$D = -1$

The discrepancy shown in Table 5.10 regarding the product $CB = D$ is sufficient to demonstrate that the representation suggested is not a true representation of the group. This way of demonstrating such a conclusion, is, in general, rather tedious, and we can test the possibility more readily by using (5.7). The character table of the point group C_{2h} is given as Table 5.11

Table 5.11. Character table of the point group C_{2h}

	I	C_2	σ_h	i
A_g	1	1	1	1
A_u	1	1	-1	-1
B_g	1	-1	-1	1
B_u	1	-1	1	-1

Then we have

	I	C_2	σ_h	i
$\chi(R)$	1	-1	1	1
g_R	1	1	1	1
$g_R\chi(R)$	1	-1	1	1

$$a_{(A_g)} = 1/4 \ [(1.1) + (-1.1) + (1.1) + (1.1)] = \frac{1}{2}$$

$$a_{(A_u)} = 1/4 \ [(1.1) + (-1.1) + (1.-1) + (1.-1)] = -\frac{1}{2}$$

$a_{(B_g)}$ $= 1/4 \ [(1.1) + (-1.-1) + (1.-1) + (1.1)] = 0$

$a_{(B_u)}$ $= 1/4 \ [(1.1) + (-1.1) + (1.1) + (1.-1)] = 0$

As we see, the total number of irreducible representations contained in the representation whose character is given above, is zero. This shows very clearly that the representation is not a true representation. However, it is not necessary for the total to be zero in order to show this; any fractional answer such as

$a_{Ag} = \dfrac{1}{2}$ or any negative answer such as $a_{A_u} = -\dfrac{1}{2}$ is sufficient to show that the proposed representation is not a true one. In practice, we will find the check by reduction very useful in showing whether we have constructed reducible representations correctly. There will be many such examples in Chapters 6 and 7.

Note that in checking by the first method, we have to multiply the actual matrices together, and not their characters. Where we have a one-dimensional representation, it consists of a single number, which is really a 1 x 1 matrix and, since the matrix has only this one element, the matrix and its character are identical. For degenerate representations (i.e. of dimension greater than 1), the matrix and its character are, of course, not the same.

PROBLEMS

1 If Q R, S, T are the following matrices:

$$Q = \begin{bmatrix} 2 & -1 \\ 3 & 1 \\ 1 & 0 \end{bmatrix} \quad R = \begin{bmatrix} 3 \\ 5 \\ 1 \end{bmatrix} \quad S = [1 \quad 0] \quad T = \begin{bmatrix} 1 & 4 & -2 \\ 2 & 0 & 1 \end{bmatrix},$$

state which of the products can be formed, and determine them.

2 Construct the reducible representation in terms of Cartesian displacement co-ordinates for a molecule $A_2 B_2$ of C_{2h} symmetry, and determine how many times each irreducible representation of the group C_{2h} occurs in it.

3 Show, by constructing a table of $X^{-1}GX$, that the symmetry operations of the point group S_4 fall into four classes.

4 With the aid of the character tables in Appendix II, reduce the following true representations of the point groups given:

(a) T_d

	I	$8C_3$	$3C_2$	$6S_4$	$6\sigma_d$
$\chi(R)$	10	1	2	0	4

(b) D_{3h}

	I	$2C_3$	$3C_2$	σ_h	$2S_3$	$3\sigma_v$
$\chi(R)$	6	0	2	2	2	2

(c) C_{3v}

	I	$2C_3$	$3\sigma_v$
$\chi(R)$	6	0	0

(d) C_{4v}

	I	$2C_4$	C_2	$2\sigma_v$	$2\sigma_d$
$\chi(R)$	4	0	0	2	0

5. By constructing the group multiplication table, and multiplying together the appropriate matrices, show whether the following representations of C_{2v} are true representations

		I	C_2	$\sigma_v(yz)$	$\sigma_v'(xz)$
(a)	Γ_a	1	-1	1	1

(b)

$$\Gamma_b = \begin{bmatrix} 1 & 0 \\ 0 & 1 \end{bmatrix} \begin{bmatrix} -\tfrac{1}{2} & 0 \\ 0 & -\tfrac{1}{2} \end{bmatrix} \begin{bmatrix} \tfrac{1}{2} & 0 \\ 0 & \tfrac{1}{2} \end{bmatrix} \begin{bmatrix} -1 & 0 \\ 0 & -1 \end{bmatrix}$$

6. By using equation (5.7), show whether Γ_b of question 5(b) is a true representation of C_{2v}, and if so, reduce it.

6

Symmetry and Physical Properties

In the previous chapters we have considered the basic symmetry elements and
their operations. We have discussed point group and space group symmetry and
have introduced the concepts of group theory. We are now in a position to consider the effects of symmetry on various physical properties of crystals and
molecules and of the implications of a knowledge of symmetry in the determination of crystal and molecular structure.

In crystal symmetry we shall see how the technique of X-ray diffraction is
used to determine the crystal class and Bravais lattice of a crystalline material.
We shall also see how the same technique can be used to give information on
the space group to which the material belongs. The importance of a knowledge
of space-group symmetry in simplifying the task of determination of crystal
structure will be explained.

The symmetry of a molecule has a profound effect on its physical properties.
The existence of dipole moment or optical activity in a molecule is compatible
only with certain types of point group, and point group symmetry also determines
the number and activity of the bands observed in the vibrational spectrum of a
molecule. Once the point group of a molecule has been determined, the arrangement of the atoms may be deduced, though not the interatomic distances.
Similarly, we may decide on pure symmetry arguments which transitions between
the energy levels of a molecule are allowed, though not which are energetically
feasible. We should therefore bear in mind that symmetry arguments, though of
great value, are in an important sense qualitative and not quantitative.

Symmetry and X-ray Crystallography

In Chapter 4 we saw that a crystalline array of asymmetric objects could be
more simply represented by replacing each asymmetric unit by a lattice point
to produce a space lattice. This simplification of a three-dimensional structure

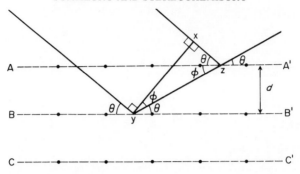

Figure 6.1   Conditions for reinforcement of diffracted X-radiation from a lattice

can also be used to simplify discussion of diffraction effects. If we consider a two-dimensional lattice, such as that shown in Figure 6.1, we can regard it as being made up of stacks of parallel planes, AA', BB', CC', etc, in which the distance between adjacent planes is d. This lattice can act as a diffraction grating for X-radiation and we can work out the conditions for reinforcement of the diffracted beams as follows:

If we consider a beam of X-radiation with a wavefront XY hitting the lattice with an incident angle θ, some of the X-rays will be reflected (diffracted) at position Z by the upper plane (AA') and some at Y by the lower plane (BB'). If we want to find out the conditions under which these reflected beams will reinforce each other we must apply the normal laws of reflection, i.e. the condition for reinforcement is that the path difference must be an integral number of wavelengths (λ) of the radiation used.

The path difference is $YZ - XZ$

$\therefore$ for reinforcement $YZ - XZ = n\lambda$

i.e. $YZ - YZ \sin\phi = n\lambda$

$\therefore$ $YZ (1 - \sin\phi) = n\lambda$

but $\phi = 90 - 2\theta$

$\therefore$ $YZ (1 - \sin [90 - 2\theta]) = n\lambda$

$\therefore$ $YZ (1 - \cos 2\theta) \qquad = n\lambda$

$\therefore$ $YZ \cdot 2\sin^2\theta \qquad = n\lambda$

but $YZ = d/\sin\theta$

$\therefore$ $2d \sin\theta = n\lambda$

This relationship $n\lambda = 2d \sin\theta$ is called *Bragg's Law,* and means that, if we have a diffraction grating made up from planes through lattice points, diffraction only occurs when the incident angle of the X-radiation has a value such that $n\lambda = 2d\sin\theta$.

X-RAY DIFFRACTION AND THE UNIT CELL

The lattice which describes a crystalline material can be divided up into stacks of parallel planes in an infinite number of ways. Figure 6.2 shows a two-dimensional lattice divided up into various stacks of parallel planes with spacings d_m each one of which will have its own diffraction condition $n\lambda = 2d_m \sin\theta_m$. These arguments are readily extended to three-dimensional lattices where again it is possible to pick out stacks of parallel planes. Each stack of parallel planes in a three-dimensional lattice represents a possible crystal face. As we have seen in Chapter 1, the faces actually appearing at the surface will be those planes on which crystal growth is slowest. We can identify the stacks of parallel planes by the Miller indices (h,k,l) in the same way as crystal faces are identified by these indices. The Miller indices of a stack of parallel planes represent the orientation of the planes with respect to the three major axes of the crystal. The

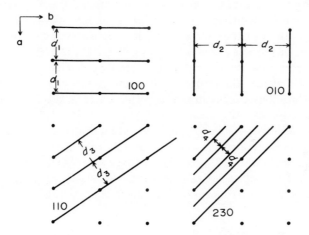

Figure 6.2   Stacks of parallel planes in a two-dimensional lattice.

indices for a given stack of planes in a lattice can be obtained by counting the number of times the planes cut the axes from one lattice point to the next, including one of the lattice points. If we assume that the lattice of Figure 6.2 is the ab plane of a crystal then all of the stacks of planes shown lie in the plane $c = 0$ and therefore all of them have $l = 0$. The planes with spacings d_1 cut the a axis at each lattice point and $h = 1$, the planes lie parallel to the b axis and $k = 0$. The Miller indices of the planes with spacing d_1 are therefore (100). Similarly those with spacing d_2 are (010). The stack of planes with spacings d_3 cuts both the a and b axes once between the lattice points and therefore has the Miller indices (110). The stack of planes with spacings d_4 has indices (230).

In a single crystal each stack of parallel planes is unique and has a definite

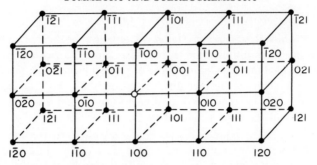

Figure 6.3   Part of a reciprocal lattice (origin at O) showing the planes which produce
                each diffraction lattice point

orientation with respect to the axes of the crystal. If the crystal is rotated about
one of its major axes and in an X-ray beam, each stack of parallel planes will, in
turn, give a spot focus diffraction beam when the Bragg's Law conditions are
obeyed. Thus, from a lattice in real space, in which each point has an identical
environment, we produce a new lattice in which the lattice points are the
diffraction spots each one of which is produced by reflection of X-radiation
from a stack of planes. This lattice of diffraction spots is called a *Reciprocal
Lattice* of which Figure 6.3 is an example. The reciprocal lattice is so called
because the reciprocal cell dimensions ($a^*, b^*, c^*, \alpha^*, \beta^*, \gamma^*$) are related to the
real cell dimensions ($a, b, c, \alpha, \beta, \gamma$) by a reciprocal relationship. Distance in
reciprocal space is proportional to the reciprocal of the distance in real space
(e.g. $a^* = k/a$) and reciprocal space angles (except for triclinic cells) are the
supplements of real cell angles, (e.g. $\alpha^* = 180 - \alpha$). Figure 6.4 shows the
relationship between two-dimensional real and reciprocal cells.

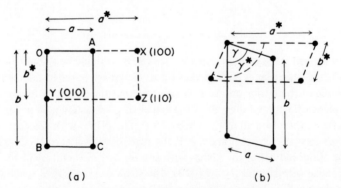

(a)   (b)

Figure 6.4   Relationship between a real and a reciprocal lattice in two dimensions. (a)
                for a rectangular system the real cell is OACB and the reciprocal cell OXZY
                (b) for an oblique system the real cell angle γ is the supplement of γ^*

The X-ray diffraction pattern of a single crystal gives us information on the size and shape of the reciprocal lattice and from this we can get the size and shape of the real cell, i.e. we can determine the crystal class. Once we know the crystal class we have reduced the possible number of space groups from 230 to the number associated with its crystal class (see Table 4.5). The proportionality constant for rectangular crystals is the wavelength (λ) of the X-radiation used and the relationship between the real and reciprocal cells is:

$$a = \lambda/a^* \qquad \alpha = 180 - \alpha^*$$
$$b = \lambda/b^* \qquad \beta = 180 - \beta^*$$
$$c = \lambda/c^* \qquad \gamma = 180 - \gamma^*$$

For example the reciprocal cell dimensions of rutile (TiO_2) obtained with Cu Kα radiation are $a^* = b^* = 0.336$ r.u. (reciprocal units). $c^* = 0.521$ r.u. $\alpha^* = \beta^* = \gamma^* = 90°$. The crystals are therefore tetragonal with

$$a = b = \frac{1.542}{0.336} = 4.59\text{Å} \quad c = \frac{1.542}{0.521} = 2.96\text{Å}$$

Once we know that a material, such as TiO_2, crystallises in the tetragonal system we have reduced the number of possible space groups from 230 to the 68 compatible with this crystal class.

The reciprocal cell dimensions for $BaSO_4$ obtained with Cu Kα radiation are $a^* = 0.215$ r.u., $b^* = 0.174$ r.u., $c^* = 0.283$ r.u. $\alpha^* = \beta^* = \gamma^* = 90°$. The crystals of $BaSO_4$ are therefore orthorhombic with $a = 7.13\text{Å}$, $b = 8.86\text{Å}$, $c = 5.41\text{Å}$ and the number of possible space groups for this system is reduced from 230 to the 59 compatible with orthorhombic symmetry.

The relationship between the real and reciprocal cells for non-rectangular systems is more complicated and, for example, for monoclinic cells with the b axis unique is

$$a = \frac{\lambda}{a^*\sin\beta} \quad b = \frac{\lambda}{b^*} \quad c = \frac{\lambda}{c^*\sin\beta}$$

$$\alpha = \alpha^* = 90° \quad \beta = 180 - \beta^* \quad \gamma = \gamma^* = 90°$$

and for hexagonal crystals with c unique

$$a = b = \frac{\lambda}{a^*\sin60} \quad c = \frac{\lambda}{c^*}$$

$$\alpha = \beta = \alpha^* = \beta^* = 90° \quad \gamma = 60° = 180 - \gamma^*.$$

The reciprocal cell dimensions of d-tartaric acid obtained with Cu Kα radiation are $a^* = 0.203$ r.u., $b^* = 0.257$ r.u., $c^* = 0.253$ r.u., $\alpha^* = \gamma^* = 90°$, $\beta^* = 79°50'$. The cell is therefore monoclinic with dimensions, $a = 7.72$Å, $b = 6.00$Å $c = 6.20$Å, $\beta = 100° 10'$. The number of possible space groups for D-tartaric acid is reduced to the 13 compatible with the monoclinic class.

X-RAY DIFFRACTION AND THE BRAVAIS LATTICE

The reciprocal lattice obtained from X-ray diffraction studies on a crystal enables us to determine the size and shape of the unit cell in real space. A study of the type of reciprocal lattice produced can also give information on the type of Bravais lattice to which the crystal belongs. The type of cell can be identified by systematic absences of reciprocal lattice points for X-ray reflections of the type $h\,k\,l$. The types of systematic absences shown by the various lattices are:

(i) If $h\,k\,l$ reciprocal lattice points are absent when $h + k + l$ is odd, the cell is body centred (I).

(ii) If $h\,k\,l$ reciprocal lattice points are absent when $h + k$ is odd, the cell is C-face centred (C). Similar conditions apply to A and B centring.

(iii) If $h\,k\,l$ reciprocal lattice points are present only when $h + k$, and $h + l$, and $k + l$ are even, the cell is face-centred on all faces (F).

(iv) If there are no systematic absences of reciprocal lattice points the cell is primitive (P).

It is important in looking for absences of reflections producing a reciprocal lattice that we distinguish between systematic absences and those due to chance zero intensity reflections. For a set of absences to be systematic all reflections of the set must be absent. The list of systematic absences given above for the Bravais lattices is general and applies to single crystal data for all crystal classes. The X-ray powder diffraction patterns of cubic crystals, however, provide a

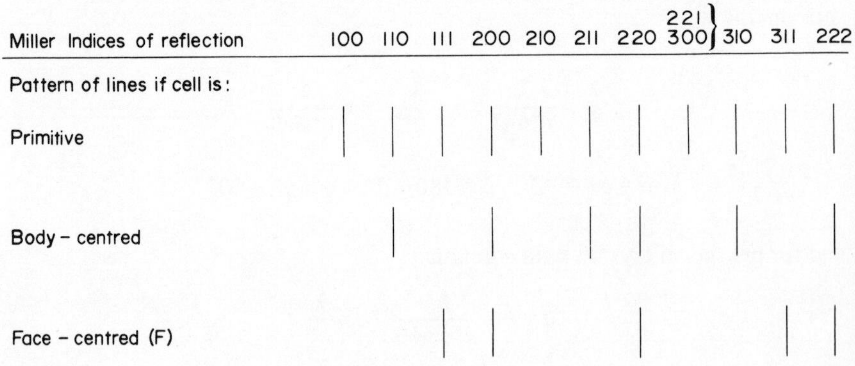

Figure 6.5   X-ray diffraction powder patterns for cubic crystals

simple example of these effects. As we can see from Figure 6.5 the powder
pattern of a primitive cubic cell contains all of the possible reflections in the
region illustrated, the body-centred pattern contains no reflections for which
$h + k + l$ is odd and the face-centred (F) pattern contains only reflections where
all binary combinations of the three indices are even.

Reference to Table 4.6 shows that a knowledge of the Bravais lattice of a
crystal can again restrict the number of possible space groups for the system. For
example the 59 possible orthorhombic space groups are reduced to 30, if the cell
is P; to 5, if the cell is F; to 9, if the cell is I and to 15 if the cell is A or C.

X-RAY DIFFRACTION AND SPACE GROUPS

In X-ray diffraction studies it is possible to obtain information on two-
dimensional sections of the reciprocal lattice. The most useful sections are the
$h\,k\,0$, $h\,0\,l$ and $0\,k\,l$ layers of the lattice because from them we can obtain
information on the symmetry about the a, b and c axes respectively. Again the
symmetry information is obtained from systematic absences of X-ray reflections.
This is illustrated in Table 6.1 which lists the absences and the possible reflections
expected in an $h\,k\,0$ projection for a number of possible symmetries about the
c axis.

So far we have made use of systematic absences without explaining why a
given type of crystal symmetry should produce absences in a reciprocal lattice.
This we can now illustrate with reference to the simplest case, the absences
produced by a screw axis, Figure 6.6 shows a real lattice which has a 2_1 axis
parallel to a as shown. The stack of parallel planes (100) will give diffraction when
$\lambda = 2a\sin\theta$ (i.e. $d_{100} = a$), that is, when reflections from the surfaces DD and FF
are one wavelength apart. The presence of the screw axis, however, means that
there is another surface (EE) $a/2$ from DD and FF. Reflections from this surface
will be exactly out of phase with those from DD and FF and will cancel them
out. This will happen for every $(h00)$ plane where h is odd. For the (200)

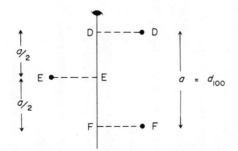

Figure 6.6   A 2-fold screw axis and systematic absences of X-ray reflections

Table 6.1. Systematic absences (and reflections present) produced by various symmetries about the c axis .

Symmetry	Absent Reflections	Present Reflections	Symbol
glide plane normal to c glide $a/2$	$h\,k\,0$ absent for h odd	$h\,k\,0$ $(h=2n)$	a
glide plane normal to c glide $b/2$	$h\,k\,0$ absent for k odd	$h\,k\,0$ $(k=2n)$	b
glide plane normal to c glide $\dfrac{a+b}{2}$	$h\,k\,0$ absent for $h+k$ odd	$h\,k\,0$ $(h+k=2n)$	n
glide plane normal to c glide $\dfrac{a+b}{4}$	$h\,k\,0$ absent unless $h+k=4n$	$h\,k\,0$ $(h+k=4n)$	d
2-fold screw axis parallel to a	$h\,0\,0$ absent for h odd	$h\,0\,0$ $(h=2n)$	2_1
3-fold screw axis parallel to a	$h\,0\,0$ absent unless $h=3n$	$h\,0\,0$ $(h=3n)$	$3_1, 3_2$
4-fold screw axis parallel to a	$h\,0\,0$ absent unless $h=4n$	$h\,0\,0$ $(h=4n)$	$4_1\,4_3$
4-fold screw axis parallel to a	$h\,0\,0$ absent for h odd	$h\,0\,0$ $(h=2n)$	4_2
6-fold screw axis parallel to a	$h\,0\,0$ absent unless $h=6n$	$h\,0\,0$ $(h=6n)$	$6_1, 6_5$
6-fold screw axis parallel to a	$h\,0\,0$ absent unless $h=3n$	$h\,0\,0$ $(h=3n)$	$6_2, 6_4$
6-fold screw axis parallel to a	$h\,0\,0$ absent for h odd	$h\,0\,0$ $(h=2n)$	6_3

(similar conditions for screw axis parallel to b).

plane, however, the reflections from DD and FF are two wavelengths apart
(i.e. $2\lambda = 2a\sin\theta$). This means that reflections from EE will differ by λ and that
they will reinforce those from DD and FF. Every reflection $h\,0\,0$ for which
h is even will likewise be present.

It would appear that if we obtain the zero layer $h\,k\,0$, $h\,0\,l$, $0\,k\,l$ lattices
about all three major axes of a crystal, we can write down the space group for
that crystal. This is true for a number of space groups. For example, if we have
the following systematic absences for an orthorhombic crystal:

$$
\begin{array}{lll}
h\,k\,l & \text{— no absences} & P \\
0\,k\,l & \text{— no absences} & \\
h\,0\,l & \text{— no absences} & \\
h\,k\,0 & \text{— no absences} & \\
h\,0\,0 & \text{— absent when } h \text{ is odd} & 2_1 \\
0\,k\,0 & \text{— absent when } k \text{ is odd} & 2_1 \\
0\,0\,l & \text{— no absences} & 2
\end{array}
$$

We have screw axes along the a and b axes as the only symmetry elements and
the space group is $P2_1 2_1 2$. This space group is one of the 70 that are uniquely
determined by systematic absences. The other 160 are not uniquely determined
because systematic absences are produced by translation symmetry effects and if
two or more space groups differ by only pure rotation or reflection symmetry
these absences cannot distinguish between them. For example the triclinic space
groups P1 and P$\bar{1}$ differ by only the centre of symmetry and we cannot distinguish
between them by systematic absences. Similarly if an orthorhombic crystal has:

$$
\begin{array}{lll}
h\,k\,l & \text{— no absences} & P \\
0\,k\,l & \text{— absent when } k+l \text{ is odd} & n \\
h\,0\,l & \text{— absent when } h \text{ is odd} & a \\
\left.\begin{array}{l} h\,k\,0 \\ h\,0\,0 \\ 0\,k\,0 \end{array}\right\} & \begin{array}{l}\text{— no additional absences not} \\ \text{covered by the absences} \\ \text{listed above}\end{array} & \\
0\,0\,l & \text{— absent when } l \text{ is odd} & 2_1
\end{array}
$$

the space group should be $Pna2_1$. This space group, however, is not uniquely
determined because it differs from Pnma only by the presence of a centre of
symmetry in Pnma.

From X-ray diffraction data we can obtain information on the size and
shape of the unit cell of a crystal, its Bravais lattice and its space group. For
some symmetries the space group is not uniquely determined but is reduced
to a small choice, usually two but in any case to a maximum choice of four.
In these cases the final space group determination must be made by other

methods such as optical microscopy, tests for the absence of a centre of symmetry (presence of pyro- or piezo-electricity or in some cases optical activity), or by statistics based on the intensities of the diffracted X-ray beams.

SPACE GROUP INFORMATION AND CRYSTAL STRUCTURE

Space group information on a crystal is obtained directly from the reciprocal lattice produced by X-ray diffraction. To obtain the positions of the atoms it is necessary to measure the intensities of all of the X-ray reflections and from these to obtain, essentially by trial and error methods, structural information. A knowledge of the space group and of the number of atoms contained in the cell is, however, very important because this simplifies and sets the size of the problem for the crystallographer.

Once we know the unit cell shape and size we can obtain the number of formula units in the cell provided we know or can obtain the density of the crystals. The number of formula units (Z) is given by the relationship

$$Z = \frac{\rho V}{1 \cdot 66M}$$

where ρ is the density in g/cm^3. V is the unit cell volume in Å^3 (Table 6.2 gives the formulae used to obtain V for all crystal classes) and M is the is the molecular weight of the formula unit.

Table 6.2. Cell volumes for the various types of cell

Cell	Volume
Cubic	a^3
Tetragonal	$a^2 b$
Orthorhombic	abc
Monoclinic	$abc\sin\beta$
Hexagonal	$abc\sin60°$
Rhombohedral	$a^3[1 - 3\cos^2\alpha + 2\cos^3\alpha]^{\frac{1}{2}}$
Triclinic	$abc[1 + 2\cos\alpha \cos\beta \cos\gamma - \cos^2\alpha - \cos^2\beta - \cos^2\gamma]^{\frac{1}{2}}$

It is not necessary to have a very accurate density measurement because Z must be a whole number and the density need only be accurate enough to indicate the value of Z.

Zinc hydroxide has an orthorhombic unit cell with $a = 8\cdot53$Å, $b = 5\cdot16$Å, $c = 4\cdot92$Å and thus has $V = 216\cdot5$Å^3. The molecular weight of the Zn(OH)$_2$ unit is $99\cdot37$ and its density is $3\cdot07$ g/cm^3. This gives $Z = 4$ and means that there are four Zn(OH)$_2$ groups in the unit cell. From the point of view of structure determination, this suggests that in order to describe the complete structure we need only determine the position of four Zn atoms, four O atoms and eight H atoms.

The space group for this material is, however, known to be $P2_1 2_1 2_1$. This space group has a 4-fold set of general equivalent positions at

$$x,y,z; \quad (\tfrac{1}{2}-x),\bar{y},(\tfrac{1}{2}+z); \quad (\tfrac{1}{2}+x),(\tfrac{1}{2}-y),\bar{z}; \quad \bar{x},(\tfrac{1}{2}+y),(\tfrac{1}{2}-z)$$

and no special positions, Since we have to fit four zinc atoms into the cell, if we know the position of one of them, the space group requirements will automatically generate the other three. The same is true of the oxygen and hydrogen atoms. This means that, because of our knowledge of space group symmetry, we have reduced the size of the problem of determining the structure of $Zn(OH)_2$ to a minimum, i.e. we need only find the positions of one Zn atom, one O atom and two H atoms to describe the structure completely.

Similarly PdS_2 has an orthorhombic cell with $a = 5.46$Å, $b = 5.54$Å, $c = 7.53$Å and contains four formula units in a cell with space group Pbca. The cell therefore contains four Pd atoms and eight S atoms. The space group has an 8-fold set of general equivalent positions at

$$x,y,z; \quad (\tfrac{1}{2}+x),(\tfrac{1}{2}-y),\bar{z}; \quad \bar{x},(\tfrac{1}{2}+y),(\tfrac{1}{2}-z); \quad (\tfrac{1}{2}-x),\bar{y}(\tfrac{1}{2}+z); \quad \bar{x},\bar{y},z;$$
$$(\tfrac{1}{2}-x),(\tfrac{1}{2}+y),z; \quad x,(\tfrac{1}{2}-y),(\tfrac{1}{2}+z); \quad (\tfrac{1}{2}+x),y(\tfrac{1}{2}-z).$$

The eight sulphur atoms lie in these equivalent positions in the cell and we only need to determine the position of one of them. Since there are four Pd atoms in the cell they must be placed in special positions. The only set of special positions in this space group are those produced by placing atoms at a centre of symmetry, e.g. at

$$0,0,0; \quad \tfrac{1}{2},\tfrac{1}{2},0; \quad 0,\tfrac{1}{2},\tfrac{1}{2}; \quad \tfrac{1}{2},0,\tfrac{1}{2}.$$

In this example a knowledge of symmetry alone fixes the positions of the Pd atoms in the cell and all that is required to complete the crystal structure is the location of one S atom.

The measurement of the positions of the diffracted X-ray beams from a single-crystal gives us the reciprocal lattice. From this we can obtain the size, shape and Bravais lattice of the unit cell (the smallest building block of the crystal) and, with a knowledge of the density, the number of formula units in the cell. Information on the space group of the system can also be obtained from absences in the zero layer planes of the reciprocal lattice. This is the maximum amount of information that can be obtained from measurements of the positions of the X-ray reflections. In order to carry out a crystal structure determination we must make use of measurements of the intensities of these reflections. In every structure determination, however, the aim is to derive the structure from as few variable parameters as possible. It would be possible to obtain the structure by locating the position of every atom in the cell independently, but this is clearly wasteful if we know the space group of the system. As we have seen, the importance of a knowledge of the space group is that it does minimise the number of atomic positions that have to be determined.

Symmetry-determined Properties of Molecules

Much information can be deduced about the physical properties of a molecule from consideration of its point group symmetry. Such deductions may be viewed as arising from the principle, due to *Neumann,* that the physical properties of a system are invariant to its symmetry operations. We cannot alter any physical property of a molecule simply by carrying out a symmetry operation on the molecule. Some of the more important physical properties of molecules to which such arguments apply are electric and magnetic dipole moments, optical activity and electronic and vibrational spectra..

ELECTRIC DIPOLE MOMENT

The existence of a permanent electric dipole moment in a molecule can be determined by a simple symmetry argument; further, this argument also gives the direction in which the dipole moment lies if it exists. The symmetry argument cannot tell us which is the positive and which the negative end of the dipole, nor can it tell us anything about its magnitude.

There are three situations in which a molecule can have no permanent dipole moment:
1. If the molecule has a centre of symmetry.
2. If the molecule has a plane of symmetry perpendicular to the principal axis.
3. If the molecule has more than one axis of rotation.

Electric dipole moment, μ is a vector and is defined as follows:
If positive and negative charges $\pm q$ are separated by a distance r, then $\mu = q\mathbf{r}$ and the magnitude of μ is the product qr. Now by Neumann's principle the vector μ must be invariant to the symmetry operations of the molecule. Therefore the line representing the vector must be coincident with every symmetry element of the molecule. If the molecule contains a centre of symmetry, the vector μ can be wholly coincident with this one point only if the magnitude of μ is zero. In the other two cases, the symmetry elements themselves are not coincident — we either have a plane perpendicular to an axis or we have more than one axis — and the vector cannot lie in these non-coincident directions simultaneously, so again its magnitude must be zero. Possession of a permanent electric dipole moment is therefore confined to those molecules with only a single rotation axis, belonging to C_n point groups, and to those in which all the symmetry elements intersect in one line. (This situation occurs only with the C_{nv} point groups, where the $n\sigma_v$ planes intersect in the C_n axis.) In both cases the direction of μ is the direction of

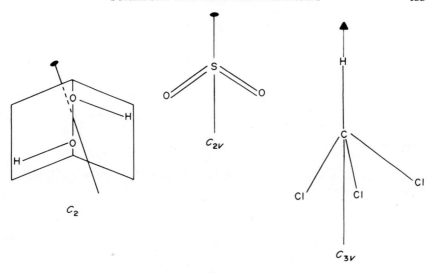

Dipole moment is directed along the C_n axis

Figure 6.7   Dipole moment and symmetry

the C_n axis. We should note that molecules belonging to S_n groups do not have an electric dipole moment; although the symmetry element is formally written as (for example) S_4, it really consists of two elements, a rotation axis and a plane perpendicular to that axis. Figure 6.7 illustrates this discussion.

OPTICAL ACTIVITY

A compound is optically active if its mirror image is not superimposable on the original. This situation occurs only in molecules which do not have a rotor-reflection axis. All other criteria which have been suggested are really more specialised versions of this condition. Optically active molecules must therefore have a point group C_n, D_n or T.

(a) The asymmetric carbon atom

A molecule containing a carbon atom with four different substituents arranged tetrahedrally may be deprived of all symmetry except for the trivial symmetry element C_1. This may be illustrated by a simple molecule of the form Cabcd such as lactic acid, or a molecule in which only some of the carbon atoms making up its skeleton are asymmetric; sugars fall into this category. Figure 6.8 shows some molecules with asymmetric atoms. All such asymmetric molecules are optically active, but not all optically active molecules are asymmetric.

Any molecule Xabcd where X is tetrahedrally co-ordinated is asymmetric; mole-

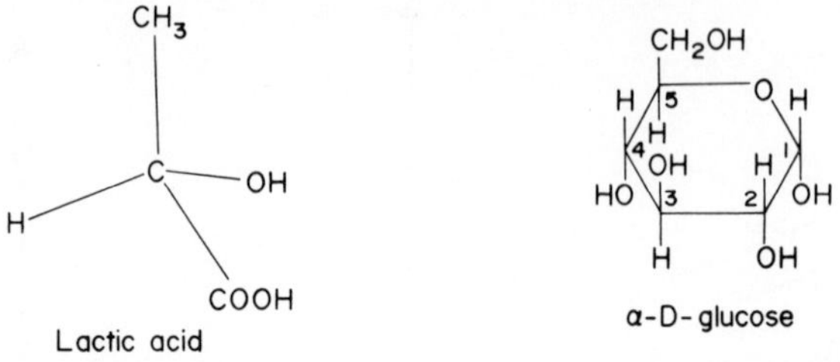

Figure 6.8   Optically active molecules which are asymmetric

cules in which X is Si, Ge, Sn, N, P and As are known. One such molecule is the organo-tin derivative:

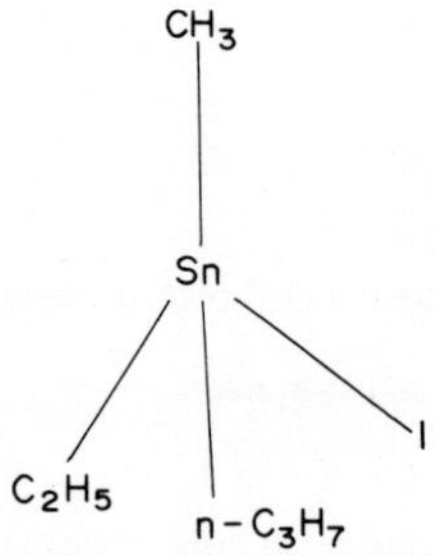

(b) Dissymmetric molecules

It is often stated, correctly, that molecules with a plane or a centre of symmetry must be optically inactive, since the possession of these symmetry elements leads to the existence of superimposable mirror images. This is another specialised version of the conditions involving rotor-reflection axes, for we have already seen that the plane of symmetry, m or σ, is equivalent to the rotor-reflection axis S_1, while the centre of symmetry, $\bar{1}$ or i, is equivalent to the rotor-reflection axis S_2.

It is possible for a molecule to retain some symmetry elements and still display optical activity. The possession of a rotation axis, n or C_n, does not lead to the existence of superimposable mirror images. Such molecules are known as dissymmetric molecules. An example is 1,3-dichloroallene, in which the only symmetry element is a 2-fold axis (see Figure 6.9). Similarly the hypothetical 'gauche' form of ethane would be optically active, although possessing four symmetry elements, a C_3 and $3C_2$, as shown in Figure 3.14.

The converse of the statement made at the beginning of this section is not true, for there are some optically inactive molecules with neither a centre nor a plane of symmetry. These are the molecules with n-fold rotor-reflection axes where n is 4 or an integral multiple of 4. The best known of these have S_4 axes; one example being the spiran illustrated in Figure 3.12.

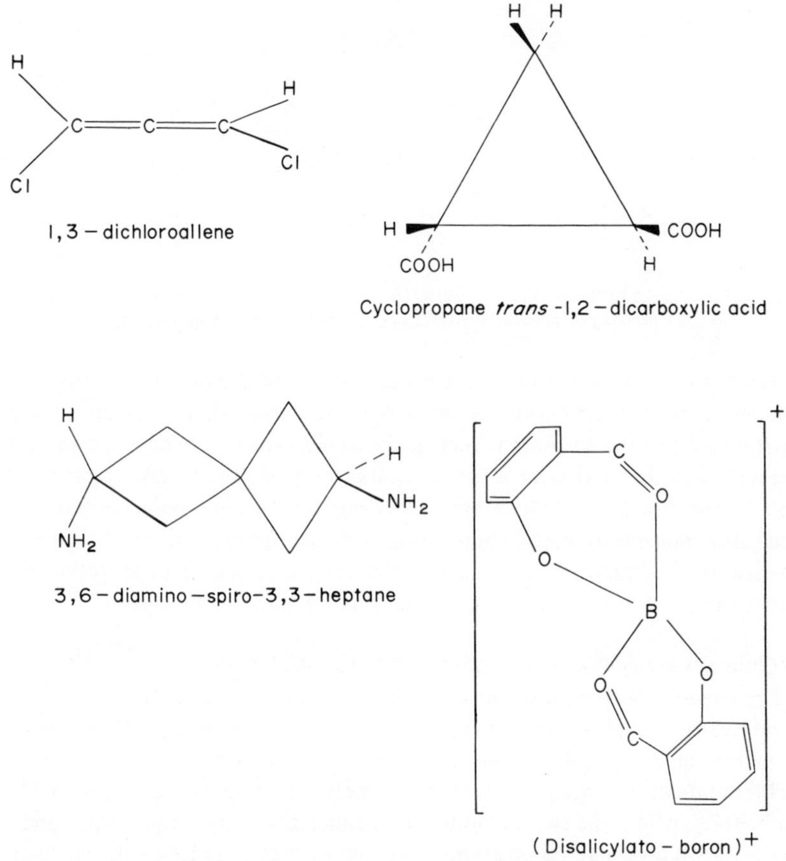

1,3 − dichloroallene

Cyclopropane *trans* -1,2 − dicarboxylic acid

3,6 − diamino − spiro-3,3 − heptane

(Disalicylato − boron)$^+$

Figure 6.9   Optically active molecules and ions which are dissymmetric

An example from inorganic chemistry of an optically active dissymmetric species based on a tetrahedral shape is the $B(\text{salicylato})_2^+$ ion.

(c) Atropoisomerism

One apparent contradiction to the general statement that the absence of an S_n axis necessarily implies optical activity has been discovered. The molecule A(= (*dextro*)-menthyl (*laevo*) menthyl-2, 6, 2′, 6′-tetranitro-4, 4′-diphenate) illustrated in Figure 6.10 is optically inactive though it possesses no symmetry elements at all. The only form in which it could have any symmetry elements is that in which the two phenyl rings are coplanar, and repulsion between the nitro groups prevents this form from being assumed. In some tetra-*ortho*-substituted biphenyls, optical activity is found because there is a large potential

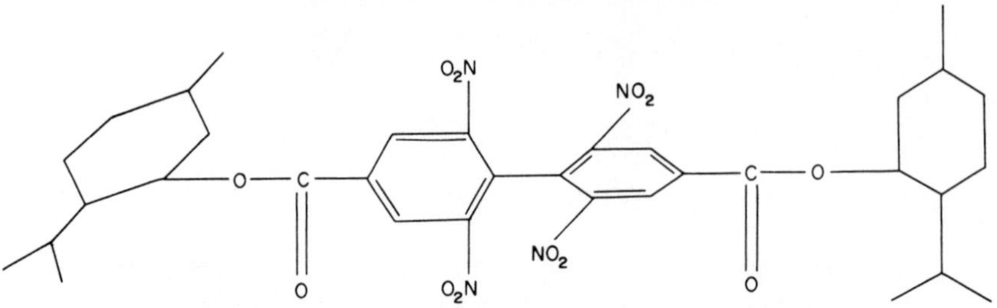

Figure 6.10   An optically inactive molecule with no symmetry elements (*dextro*)-
menthyl (*laevo*)-menthyl-2,6,2',6'-tetra nitro-4, 4'-diphenate)

energy barrier to rotation about the central bond and the two optical isomers are
isolatable; this is the phenomenon of atropoisomerism. Where, however, there
is virtually free rotation so that very many conformations are stable, every molecule with a conformation such that it rotates the plane of polarised light by some
angle in one direction will have corresponding to it another molecule causing an
equal and opposite rotation. Thus, on the average, the substance will be optically
inactive. In the case of the molecule A, the free rotation giving rise to the inactivity
is about the 4, 4' bonds linking the carboxylate groups to the rings.

(d) Optically active species based on non-tetrahedral shapes

The stereochemistry of octahedral complexes has in a number of cases been
elucidated by studies of optical activity. Two common arrangements leading to
the formation of optically active species are MB_3, where M is a metal atom and
B a bidentate ligand of C_{2v} symmetry spanning *cis* positions (see Figure 6.11)
and MB_2L_2, where the two monodentate ligands L occupy *cis*-positions, and the
two ligands B also span *cis*-positions. The first of these arrangements has the point
group D_2; one well-known ion with this symmetry is $[Co(ethylenediamine)_3]^{3+}$.
If the symmetry of B is reduced (say to C_1 as in the glycinate ligand

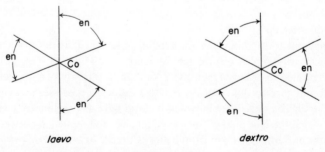

laevo   *dextro*

Figure 6.11   Optical isomers of $Co(en)_3^{3+}$

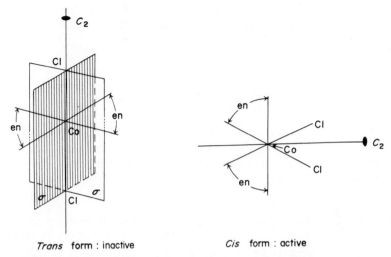

Trans form : inactive   Cis form : active

Figure 6.12   Optically active and inactive isomers of $[Co(en)_2Cl_2]^+$

$NH_2CH_2COO^-$) the overall symmetry of the complex will also be reduced and the molecule or ion will still be optically active. As an example of the second arrangement we may consider cis-$[Co(ethylenediamine)_2Cl_2]^+$, which has C_2 symmetry. Here we may distinguish between the cis- and trans-forms, since the trans-form has two planes of symmetry bisecting the equatorial plane and intersecting on the C_2 axis along the Cl—Co—Cl direction and is inactive. Figure 6.12 illustrates this.

Normally an arrangement based on a square-planar skeleton will be inactive, since the molecular plane is a plane of symmetry, but, by appropriate substitution, a molecule with no plane of symmetry may be produced. The complex illustrated in Figure 6.13 was prepared in order to demonstrate that the arrangement of the bonds round Pt was not tetrahedral. If the bonds around the Pt were tetrahedrally disposed, the molecule would possess a plane of symmetry (the plane of the ring containing the CH_3 groups) and hence exhibit no optical activity. It is true that

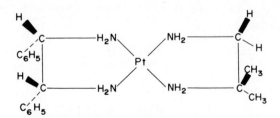

Figure 6.13   An optically active complex in which the central atom makes four coplanar bonds with the ligands

optical activity would be shown if the arrangement around the Pt were either square-planar or tetragonal-pyramidal, but studies of other complexes in fact eliminated the second possibility.

VIBRATIONAL SPECTRA

The study of vibrational spectra in connection with molecular symmetry yields the answers to three questions, namely:

1. How many vibrations of the molecule belong to the each of the irreducible representations (symmetry classes) of the molecular point group?
2. Which of the vibrations display infra-red or Raman activity?
3. What is the direction of polarisation of a given vibration?

The motions of a vibrating molecule may be described in an infinite number of ways, but the most convenient description is to resolve them into a set of modes of motion called the normal modes. In a given normal mode, the atoms all move with the same frequency and in phase. It is simple to construct a reducible representation for the normal modes and the representation is then reduced using

$$a_j = \frac{1}{h}\sum_R g_R \chi(R)\chi_j(R) \qquad (6.1)$$

which gives the number of times each irreducible representation appears in the reducible representation and consequently tells us how many vibrations belong to each symmetry class.

Let us call the reducible representation for the normal modes of a molecule Γ_0. Note that this includes all the modes of motion, that is, translational, rotational and vibrational modes. The character χ_0 of this representation is simply $N_R\chi_R$, where N_R is the number of atoms left unshifted by the operation R, and χ_R is the character of the operation.

To find the character of an operation we require to know whether it is a rotation only (a so-called proper rotation) or a rotation combined with a reflection (an improper rotation).

We showed earlier that a line OP terminating at $P(x,y)$ rotated through an angle θ gave rise to a new line OP′ where the co-ordinates (x',y') of P′ were related to (x,y) by

$$x' = x\cos\theta + y\sin\theta$$
$$y' = -x\sin\theta + y\cos\theta$$

which we express in matrix form as

$$\begin{bmatrix} x' \\ y' \end{bmatrix} = \begin{bmatrix} \cos\theta & \sin\theta \\ -\sin\theta & \cos\theta \end{bmatrix} \begin{bmatrix} x \\ y \end{bmatrix}$$

the character of the matrix being $2\cos\theta$. Now if we consider a general point P (x,y,z) and rotate OP about the z axis to give OP′ where P′ has the co-ordinates (x', y', z'), we shall have

$$\begin{bmatrix} x' \\ y' \\ z' \end{bmatrix} = \begin{bmatrix} \cos\theta & \sin\theta & 0 \\ -\sin\theta & \cos\theta & 0 \\ 0 & 0 & 1 \end{bmatrix} \begin{bmatrix} x \\ y \\ z \end{bmatrix}$$

and the character of the matrix for rotation through an angle θ is $1 + 2\cos\theta$
If the rotation is combined with reflection at the (xy) plane, then the transformation will involve a change from z to $-z$, and the matrix will be

$$\begin{bmatrix} x' \\ y' \\ z' \end{bmatrix} = \begin{bmatrix} \cos\theta & \sin\theta & 0 \\ -\sin\theta & \cos\theta & 0 \\ 0 & 0 & -1 \end{bmatrix} \begin{bmatrix} x \\ y \\ z \end{bmatrix}$$

and its character is $-1 + 2\cos\theta$. We may summarise this by writing $\chi_{(R)} = \pm 1 + 2\cos\theta$ where R involves a rotation through an angle θ $(= 360°/n$ for a C_n or S_n operation) and the $\pm$ signs refer to proper (C_n) and improper (S_n) rotations respectively.

The following special cases are important:
For the identity we have $x' = x$, $y' = y$, $z' = z$. Thus the matrix is

$$\begin{bmatrix} x' \\ y' \\ z' \end{bmatrix} = \begin{bmatrix} 1 & 0 & 0 \\ 0 & 1 & 0 \\ 0 & 0 & 1 \end{bmatrix} \begin{bmatrix} x \\ y \\ z \end{bmatrix}$$

Its character is $+3$; by comparison with the general form we see that $\cos\theta = 1$, $\sin\theta = 0$. $1 + 2\cos 0° = 1 + 2 = 3$ so the identity operation corresponds to a rotation through $0°$.

For the inversion operation, $x' = -x$, $y' = -y$, $z' = -z$. In this case, each of the diagonal elements is -1 and all other elements are zero. By comparison with the general form $\cos\theta = -1$, $\sin\theta = 0$, so $\theta = 180°$, corresponding to the designation of the inversion as S_2.

For a reflection only, $x' = x$, $y' = y$, $z' = -z$. We therefore have

$$\begin{bmatrix} x' \\ y' \\ z' \end{bmatrix} = \begin{bmatrix} 1 & 0 & 0 \\ 0 & 1 & 0 \\ 0 & 0 & -1 \end{bmatrix} \begin{bmatrix} x \\ y \\ z \end{bmatrix}$$

with character 1. In this case we have $\theta = 0°$ and since the reflection is an improper operation we expect its character to be $-1 + 2\cos\theta = -1 + 2 = +1$, as shown. This corresponds to the designation of the reflection as S_1.

Thus we can write $\chi_0 = N_R\chi_R = (\pm1 + 2\cos\theta)$. From this we have to take out those parts of the representation, Γ_{trans} corresponding to the translational and Γ_{rot} to the rotational motion of the molecule. The character χ_{trans} of Γ_{trans} is given by $\pm1 + 2\cos\theta$, and the character χ_{rot} of Γ_{rot} by $1 \pm 2\cos\theta$ ($= \pm\chi_{trans}$). Thus $\chi_{vib} = \chi_0 - \chi_{trans} - \chi_{rot}$, and then we can find out how many times each irreducible representation of the molecular point group appears in Γ_{vib}.

Let us take BF_3 as an example. By following the sequence given in Chapter 3 we may see that BF_3 belongs to the point group D_{3h}. We next need to determine N_R for each symmetry operation. N_R will usually* be the number of atoms lying on the corresponding symmetry element; if, for example, an atom lies on an axis of rotation it is unshifted by a rotation about that axis; similarly, if it lies on a plane of symmetry, it is unshifted by reflection in that plane.

The symmetry operations of the group D_{3h} are given in the character table as I, $2C_3$, $3C_2$, σ_h, $2S_3$, $3\sigma_v$. For the identity, N_R is of course equal to the number of atoms in the molecule, and in this case is 4. The C_3 axis (which is also the S_3 axis) is the axis through the B atom perpendicular to the molecular plane, and only one atom (the B atom) is therefore invariant to the corresponding operations. All four atoms are invariant to σ_h, which is by definition a plane perpendicular to the principal axis, and is thus the molecular plane. One B and one F lie on each of the σ_v and thus on each of the C_2 formed by the intersection of σ_v with σ_h. We can now set up a calculation table (Table 6.3) showing χ_R and N_R for each operation, and giving χ_0, χ_{trans}, χ_{rot} and χ_{vib}.

Table 6.3. Calculation table for determining χ_{vib} of BF_3 with D_{3h} symmetry

R	Proper operations			Improper operations		
	I	$2C_3$	$3C_2$	σ_h	$2S_3$	$3\sigma_v$
N_R	4	1	2	4	1	2
$\chi_R = \pm1 + 2\cos\theta$	3	0	-1	1	-2	1
$\chi_0 = N_R\chi_R$	12	0	-2	4	-2	2
$\chi_{trans} = \chi_R$	3	0	-1	1	-2	1
$\chi_{rot} = \pm\chi_{trans}$	3	0	-1	-1	2	-1
$\chi_{vib} = \chi_0 - \chi_{trans} - \chi_{rot}$	6	0	0	4	-2	2

*This statement is true for all symmetry elements except S_n. Since S_n symmetry elements include both a rotation axis and a horizontal mirror plane, an atom must lie on both the axis and the plane to be invarient to S_n

Table 6.4. Characters of the irreducible representations of D_{3h}

	I	$2C_3$	$3C_2$	σ_h	$2S_3$	$3\sigma_v$
A_1'	1	1	1	1	1	1
A_2'	1	1	-1	1	1	-1
E'	2	-1	0	2	-1	0
A_1''	1	1	1	-1	-1	-1
A_2''	1	1	-1	-1	-1	1
E''	2	-1	0	-2	1	0

In order to reduce χ_{vib} we need the characters of the irreducible representations of the point group D_{3h}; these are given as Table 6.4.

In reducing χ_{vib} according to equation (6.1) above, it is convenient to begin by forming the product $g_R\chi(R)$. Note that χ_{vib} is now the reducible representation which is in general called $\chi(R)$ in the equation. Table 6.3 gives $\chi(R)$ and Table 6.4 gives g_R so we have from these

R	I	$2C_3$	$3C_2$	σ_h	$2S_3$	$3\sigma_v$
$g_R\chi(R)$	6	0	0	4	-4	6

The order of the group, which is the number of symmetry operations, is 12. Then from (6.1)

$$a(A_1') = \frac{1}{12}[6.1 + 0.1 + 0.1 + 4.1 + (-4.1) + 6.1] = 1$$

$$a(A_2') = \frac{1}{12}[6.1 + 0.1 + 0(-1) + 4.1 + (-4.1) + 6(-1)] = 0$$

$$a(E') = \frac{1}{12}[6.2 + 0(-1) + 0.0 + 4.2 + (-4)(-1) + 6.0] = 2$$

$$a(A_1'') = \frac{1}{12}[6.1 + 0.1 + 0.1 + 4(-1) + (-4)(-1) + 6(-1)] = 0$$

$$a(A_2'') = \frac{1}{12}[6.1 + 0.1 + 0(-1) + 4(-1) + (-4)(-1) + 6.1] = 1$$

$$a(E'') = \frac{1}{12}[6.2 + 0(-1) + 0.0 + 4(-2) + (-4.1) + 6.0] = 0$$

Thus for BF_3 with D_{3h} symmetry, $\Gamma_{vib} = 1A_1' + 2E' + 1A_2''$; there is one vibration of A_1' symmetry, one of A_2'' and 2 of E'. We may check this to some extent if we recollect that a four-atomic non-linear molecule has $(4 \times 3 - 6) = 6$ modes of vibration. Since the E' vibrations are doubly degenerate, they count as 2 each and the total number calculated is indeed 6.

The infra-red or Raman activity of a vibration may be found from the character table. In addition to containing the characters of the irreducible representations of a point group, character tables also contain three sets of symbols:

1. The symbols R_x, R_y, R_z refer to the irreducible representations to which the rotational modes of motion of the molecule belong. We shall not discuss these further as they are not relevant to the problem.

2. The symbols x,y,z or M_x, M_y, M_z or T_x, T_y, T_z. These are all equivalent: the symbols x,y,z designate those irreducible representations which transform in the same way as the co-ordinates of a general point (x,y,z) under the symmetry operations of a point group. Both the components of the dipole moment operator (M) and the translation operator (T) transform in the same way as the co-ordinates of a general point. The vibrations belonging to any irreducible representation transforming in the same way as any of these co-ordinates are infra-red active. We shall explain and illustrate this point below.

3. Binary combinations of the co-ordinates, such as x^2, y^2, z^2, xy, yz, zx and linear combinations of these. These may appear on their own or as subscripts to the symbol α. Vibrations belonging to irreducible representations transforming in the same way as these binary combinations are Raman-active. The symbol α designates the components of the polarisability tensor. For a vibration to be Raman-active, it must involve a change in polarisability and at least one of the components of α must be non-zero.

One sees in many books the relation $\mu = \alpha F$, where μ is the dipole moment induced in a molecule by the application of a field F. Both the dipole moment and the field are vector quantities, and the polarisability α is a scalar quantity only when the directions of μ and F are parallel. In general, μ and F are not parallel and each component of μ is related to every component of F, as follows

$$\mu_x = \alpha_{xx}F_x + \alpha_{xy}F_y + \alpha_{xz}F_z$$
$$\mu_y = \alpha_{yx}F_x + \alpha_{yy}F_y + \alpha_{yz}F_z$$
$$\mu_z = \alpha_{zx}F_x + \alpha_{zy}F_y + \alpha_{zz}F_z$$

The nine components of α constitute a second-rank tensor. Since it is symmetric ($\alpha_{ij} = \alpha_{ji}$) there are only six independent components. The simple relation $\mu = \alpha F$ is true when $\alpha_{xx} = \alpha_{yy} = \alpha_{zz} = \alpha$, and all other components are zero.

We said earlier that irreducible representations transform in the same way as co-ordinates or combinations of co-ordinates. Let us discuss this for the point group C_{2h}. Under the four symmetry operations of this group we have

$$(x,y,z) \xrightarrow{I} (x,y,z)$$

$$(x,y,z) \xrightarrow{C_2(z)} (-x, -y, z)$$

$$(x,y,z) \xrightarrow{\sigma(xy)} (x, y, -z)$$

$$(x,y,z) \xrightarrow{i} (-x, -y, -z)$$

where $C_2(z)$ and $\sigma_h(xy)$ denote that the C_2 axis is taken in the z direction and the plane of symmetry is the (xy) plane. If we write down the four matrices representing these transformations, we have

$$
\begin{array}{cccc}
I & C_2(z) & \sigma_h(xy) & i \\[4pt]
\begin{bmatrix} 1 & 0 & 0 \\ 0 & 1 & 0 \\ 0 & 0 & 1 \end{bmatrix} &
\begin{bmatrix} -1 & 0 & 0 \\ 0 & -1 & 0 \\ 0 & 0 & 1 \end{bmatrix} &
\begin{bmatrix} 1 & 0 & 0 \\ 0 & 1 & 0 \\ 0 & 0 & -1 \end{bmatrix} &
\begin{bmatrix} -1 & 0 & 0 \\ 0 & -1 & 0 \\ 0 & 0 & -1 \end{bmatrix}
\end{array}
$$

If we now tabulate the matrix elements involving x, y and z for each operation, we have

	I	C_2	σ_h	i
x	1	−1	1	−1
y	1	−1	1	−1
z	1	1	−1	−1

Comparing these with the characters of the irreducible representations of C_{2h} given in Table 5.11, we see that x and y have the same coefficients as the representation B_u, while z has the same coefficients as A_u. This is what we mean by saying that a co-ordinate transforms in the same way as an irreducible representation. In order to find out how the binary combinations transform, we can, for a non-degenerate point group such as C_{2h}, form the so-called direct product of the representations of the single co-ordinates; for instance, if we want to know how (xz) transforms, we multiply together, for each operation, the character for x and the character for z. If we carry this out for all six binary combinations, we obtain

	I	C_2	σ_h	i
x^2	1	1	1	1
y^2	1	1	1	1
z^2	1	1	1	1
xy	1	1	1	1
yz	1	−1	−1	1
zx	1	−1	−1	1

Comparison with the character table shows that the first four combinations transform as A_g and the other two as B_g. Thus the complete character table including the labelling of irreducible representations which transform as the angular momentum components R_x, R_y, R_z reads

Character table for the point group C_{2h}

	I	C_2	σ_h	i		
A_g	1	1	1	1	R_z	$x^2, y^2, z^2, xy, x^2-y^2$
A_u	1	1	-1	-1		z
B_g	1	-1	-1	1	R_x, R_y	yz, zx
B_u	1	-1	1	-1		x, y

Thus vibrations of the classes A_g and B_g are Raman-active, while those of the classes A_u and B_u are infra-red active. Note that no class contains vibrations which are both infra-red and Raman-active. This is true of all point groups containing a centre of symmetry, and illustrates the well-known rule of *Mutual Exclusion*. This rule states that, for a molecule with a centre of symmetry, no vibration can be both infra-red and Raman-active. We can see how this separation comes about for the group C_{2h}; the same argument applies to any centro-symmetric point group. If any one of the co-ordinates is inverted through the centre of symmetry, it changes sign; thus each of x, y, z is antisymmetric to inversion and is, by definition, of type u. On the other hand, all binary combinations involve two changes of sign and this double change is symmetric to inversion; thus binary combinations are of type g. Since infra-red active vibrations belong to representations transforming as single co-ordinates, they always belong to u representations. Raman-active vibrations belong to representations transforming as binary combinations, which are always g. Thus no vibration of a molecule belonging to a centro-symmetric group can be both infra-red and Raman-active.

An alternative method of arriving at the infra-red or Raman activity of molecular vibrations is to obtain the characters of the reducible representations Γ_μ for dipole moment and Γ_α for polarisability, and to determine which irreducible representations they contain. Any irreducible representation contained in Γ_μ corresponds to infra-red active vibrations, and any irreducible representation contained in Γ_α corresponds to Raman-active vibrations. The character, χ_μ of Γ_μ is the same as χ_{trans}, namely, $\pm 1 + 2\cos\theta$. The character, χ_α of Γ_α is $2\cos\theta (\pm 1 + 2\cos\theta)$, so these are easily obtained. Consider the point group D_{3h} as an example. We have in our previous calculation (Table 6.3) already obtained χ_{trans}. A further table (Table 6.5) can now be set up.

Table 6.5. Calculation table to determine χ_μ and χ_α for the point group D_{3h}

Operation	I	$2C_3$	$3C_2$	σ_h	$2S_3$	$3\sigma_v$	
$\pm 1 + 2\cos\theta$	3	0	-1	1	-2	1	$= \chi_\mu$
$2\cos\theta$	2	-1	-2	2	-1	2	
Product	6	0	2	2	2	2	$= \chi_\alpha$

It is left for the reader to show, by reduction of the reducible representations, that $\Gamma_\mu = 1A_2'' + 1E'$, while $\Gamma_\alpha = 2A_1' + 1E'$. Note that the numbers of times each irreducible representation appears in Γ_μ or Γ_α is *not* the number of vibrations in the symmetry class. It is in fact the number of components of the dipole moment operator **M** or polarisability tensor α which transform as the given irreducible representation (or, in the case of degenerate representations, the number of sets of such components). For D_{3h}, we may summarise this information as Table 6.6.

Table 6.6. Infra-red and Raman activity of vibrations under D_{3h} symmetry

Symmetry class	Infra-red activity	Raman activity
A_1'	inactive	active $x^2 + y^2$, z^2
A_2'	inactive	inactive
E'	active (x,y)	active $(x^2 - y^2, xy)$
A_1''	inactive	inactive
A_2''	active z	inactive
E''	inactive	inactive

For completeness the components of **M** and α transforming as the various irreducible representations are shown, though it must be emphasised that the reduction of Γ_μ and Γ_α tells us only how many such components there are but does not identify them. Components in brackets are degenerate pairs. Note that one component of **M** transforms as A_2'' and one degenerate pair as E', corresponding to $\Gamma_\mu = 1A_2'' + 1E'$, while two separate components of α transform as A_1' and one degenerate pair as E', corresponding to $\Gamma_\alpha = 2A_1' + 1E'$.

Combining this information with the previous calculation which showed the vibrations of BF_3 with D_{3h} symmetry to be $1A_1' + 1A_2'' + 2E'$, we see that the infra-red spectrum of BF_3 should contain three bands $(1A_2'' + 2E')$ and the Raman spectrum three $(1A_1' + 2E')$. There are two coincidences − that is, bands appearing in both the infra-red and Raman spectra.

When one wishes to use vibrational spectroscopy in determining molecular symmetry, the procedure is to work out, for each possible point group of the molecule, the total number of vibrations, and the numbers active in the infra-red, in the Raman, and in both (coincidences) and compare the observed spectra with the predictions in each case. Some recent investigations of this nature include the assignment of C_{2v} symmetry to SF_4, the demonstration of the ionic nature of $TeCl_4$ in the solid, where it is $TeCl_3^+ Cl^-$ (the $TeCl_3^+$ is pyramidal), and the discovery that B_2Cl_4 is non-planar with D_2 symmetry in the liquid and vapour, in contrast to its crystal structure, where X-ray diffraction measurements have shown it to be planar with D_{2d} symmetry.

Selection Rules and Polarization

Consider a transition involving a change in some property, the property being associated with an operator X. If the wave functions associated with the initial and final states of the system are ϕ_i, ϕ_f respectively, then the probability of the transition occurring is proportional to the transition moment, which is given by the integral $\int \phi_i X \phi_f d\tau$.

If we are discussing electric dipole transitions, such as we observe in electronic and vibrational spectra, then the operator X will be the dipole moment operator **M** which is a vector with components M_x, M_y, M_z. The probabilities of a transition occurring with the electric vector of the incident radiation parallel to the x, y or z directions will thus be given by

$$\int \phi_i M_x \phi_f d\tau, \quad \int \phi_i M_y \phi_f d\tau, \quad \int \phi_i M_z \phi_f d\tau,$$

respectively.

At least one of these integrals must be non-zero in order for the transition to occur, and such an integral can be non-zero only if it is, or contains a component which is, symmetric to all the operations of the group. This condition arises because the transition moment, being a physical property of the system, must be invariant to all the symmetry operations of the group, (i.e. totally symmetric).

The question of whether the integral is totally symmetric is answered by constructing the character of the reducible representation formed by multiplying together the characters of the irreducible representations to which ϕ_i, ϕ_f and the dipole moment component belong. In this connection we should recall that M_x, M_y, M_z transform like x, y, z respectively.

If we are considering transitions from the vibrational ground state to an excited state, the problem is somewhat simplified because the vibrational ground state always has a totally symmetric wave function. The character of every operation in the totally symmetric irreducible representation is +1, so we need only form the products of the characters of the irreducible representations to which the co-ordinates and the final-state wave functions belong.

Consider a molecule of C_{2h} symmetry; the totally symmetric representation is A_g. We wish to investigate the possibilities of electric dipole transitions to excited vibrational states of classes A_g, A_u, B_g, B_u. Now we have shown previously that x and y transform as B_u while z transforms as A_u. From the character table of C_{2h} we can work out all the direct products and find out whether they are, or contain, the A_g representation. If so, the transition is allowed with the electric vector of the incident radiation parallel to that direction.

The only direct products which involve the A_g representation are those resulting from a transition to an A_u state with the electric vector of the radiation parallel to z, and from a transition to a B_u state with the electric vector parallel to x or y.

When we carry out a study of the vibrational spectrum of a single crystal or an

Table 6.7. Direct products of irreducible representations of dipole moment components and excited states for C_{2h} symmetry

class of excited state	Direction	
	$x,y(B_u)$	$z(A_u)$
A_g	B_u	A_u
A_u	B_g	A_g
B_g	A_u	B_u
B_u	A_g	B_g

oriented film, we can arrange the sample in a known orientation with respect to the incident radiation and thus identify the symmetry classes to which the observed vibrations belong. Exactly the same analysis can be carried out for transitions involving a change in polarisability in order to determine the excited states accessible from the ground state by Raman scattering; Table 6.8 shows these possibilities.

Table 6.8. Direct products of the characters of the irreducible representations of polarisability components and excited states for C_{2h} symmetry

Excited states	Component	
	x^2, y^2, z^2, xy A_g	yz, xz B_g
A_g	A_g	B_g
A_u	A_u	B_u
B_g	B_g	A_g
B_u	B_u	A_u

Here we see that we can reach the A_g excited state if any of the components x^2, y^2, z^2 or xy of the polarisability tensor is non-zero and the B_g excited state if yz or xz is non-zero. Again it is possible to orient samples so as to investigate each component in turn and identify the symmetry class of each vibration.

In studying transitions where the initial state is not totally symmetric we need to form the product χ (initial state) $\times \chi$ (operator component) $\times \chi$ (final state). This occurs in vibrational spectra only when the initial state is an excited state, and is not a common situation. In electronic spectra, however, this is the usual situation, since the wave function of the ground state is very often non-totally symmetric. Consider the electronic configuration shown in Figure 6.14 belonging to a molecule of C_{2v} symmetry. We wish to establish whether electric dipole transitions are possible from either ϕ_3 or ϕ_4 to either ϕ_5 or ϕ_6.

We need to form the direct products χ(initial state) $\times \chi$(dipole moment component) $\times \chi$(final state) and see if any of these belong to or contain the totally symmetric representation of the group C_{2v}; that is the A_1 representation. In

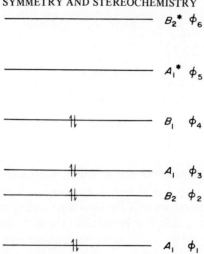

Figure 6.14   Possible electronic configuration of a molecule of C_{2v} symmetry (starred levels are antibonding)

Table 6.9 we have the necessary information, and the results are given in Table 6.10. It is convenient to begin by forming χ(initial state) $\times \chi$(final state); as the characters are all numbers, the order in which they are multiplied is immaterial.

Table 6.9. Character table of the point group C_{2v}

	I	C_2	$\sigma_v(yz)$	$\sigma_v(xz)$		
A_1	1	1	1	1	z	x^2, y^2, z^2
A_2	1	1	−1	−1		xy
B_1	1	−1	−1	1	x	xz
B_2	1	−1	1	−1	y	yz

Table 6.10. Symmetry classes of the direct products of the irreducible representations of $\phi_i \times \phi_f \times$ co-ordinate

Product	ϕ_i ϕ_f	A_1 A_1^* A_1	B_1 A_1^* B_1	A_1 B_2^* B_2	B_1 B_2^* A_2
Component					
$x(B_1)$		B_1	A_1	A_2	B_2
$y(B_2)$		B_2	A_2	A_1	B_1
$z(A_1)$		A_1	B_1	B_2	A_2

The totally symmetric representation is given by the products $(A_1 A_1^* z)$, $(B_1 A_1^* x)$ and $(A_1 B_2^* y)$.

Thus we may have electric dipole transitions from $\phi_3(A_1)$ to $\phi_5(A_1^*)$; $\phi_4(B_1)$

to ϕ_5 (A_1^*) and ϕ_3 (A_1) to ϕ_6(B_2^*) with radiation whose electric vectors are parallel to z, x and y respectively. No transition between ϕ_4 (B_1) and ϕ_6 (B_2^*) is possible.

We should bear in mind that these selection rules are deduced purely on symmetry grounds and that they tell us nothing about whether the process under consideration is energetically feasible. Similarly in the earlier discussion of optical activity and dipole moment we found that it is possible to discuss only whether or not the molecule possesses these properties, and not their magnitude or sense.

PROBLEMS

1. A lattice contains stacks of parallel planes with the d-spacings listed in the table below. CuK$_\alpha$ radiation (λ = 1·542Å) is passed through the lattice at the incident angles (θ) shown. At which values of θ do each of the stacks of parallel planes give a diffracted beam?

$$\theta° = 5 \quad 10 \quad 14 \quad 17 \quad 20 \quad 24 \quad 36$$
$$d\text{-spacings (Å). } 8·88 \quad 4·44 \quad 3·19 \quad 2·34 \quad 2·25 \quad 1·89 \quad 1·31 \quad 1·00$$

2. What are the Miller indices of the stacks of parallel planes in the plane lattice shown below if this lattice represents the ac plane of a crystalline material?

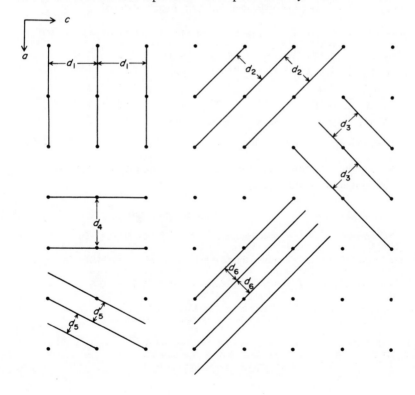

3. Given the reciprocal cell dimensions obtained with CuK_α radiation ($\lambda = 1.542$Å) for the following materials, what are their real unit cell dimensions and to which crystal class do they belong?

(i) Mo, $a^* = b^* = c^* = 0.371$ r.u.

$\alpha^* = \beta^* = \gamma^* = 90°$

(ii) Sucrose $a^* = 0.145$ r.u., $b^* = 0.177$ r.u., $c^* = 0.204$ r.u.,

$\alpha^* = \gamma^* = 90°$, $\beta^* = 77°$

(iii) SnS $a^* = 0.387$ r.u., $b^* = 0.355$ r.u., $c^* = 0.137$ r.u.,

$\alpha^* = \beta^* = \gamma^* = 90°$

(iv) $Sm(OH)_3$ $a^* = b^* = 0.282$ r.u., $c^* = 0.428$ r.u.,

$\alpha^* = \beta^* = 90°$, $\gamma^* = 120°$

(v) urea $a^* = b^* = 0.273$ r.u., $c^* = 0.328$ r.u.,

$\alpha^* = \beta^* = \gamma^* = 90°$

4. The d-spacings of cubic powder lines are given by the following formula

$$d_{hkl} = \frac{a}{(h^2 + k^2 + l^2)^{\frac{1}{2}}}$$

Given the value of a and the first five observed powder lines for the following materials, deduce the type of Bravais lattice for each material.

material		a(Å)	d-spacing of first five observed lines (Å)				
(i)	SnI_4	12.273	7.09	5.49	5.01	3.54	3.40
(ii)	CdF_2	5.388	3.11	2.61	1.90	1.62	1.56
(iii)	$PaPd_3$	4.014	4.01	2.84	2.32	2.01	1.80
(iv)	α-Sn	6.489	3.75	2.29	1.96	1.62	1.49
(v)	Y_2O_3	10.604	4.34	3.06	2.65	2.50	2.37

5. What are the cell contents in formula units for the following materials with given cell dimensions and density.

material	cell dimensions (lengths in Å)	density (g/cm^3)
(a) nitroguanidine ($CH_4N_4O_2$)	orthorhombic $a = 17.58$ $b = 24.84$ $c = 3.58$	1.78
(b) p-nitrophenol ($C_6H_5NO_3$)	monoclinic $a = 6.17$ $b = 8.95$, $c = 11.74$, $\beta = 103°$	1.46
(c) α-resorcinol ($C_6H_6O_2$)	orthorhombic $a = 10.53$ $b = 9.53$ $c = 5.66$	1.28
(d) SnI_4	cubic $a = 12.273$	4.50
(e) $(CH_3C_6H_4NC)_4CoI_2$	tetragonal $a = 14.45$ $c = 16.02$	1.55
(f) TiS (high temperature form)	rhombohedral $a = 9.04$ $\alpha = 21°48'$	4.50
(g) covellite (CuS)	hexagonal $a = 3.796$ $c = 16.36$	4.66
(h) $Ni(H_2O)_6(SnF_3)_2$	triclinic $a = 6.55$ $b = 6.73$ $c = 6.50$ $\alpha = 96.0°$ $\beta = 104.4°$ $\gamma = 98.1°$	3.30

6. The following compounds from question 5 above have as their only systematic absences of X-ray reflection:

(a) nitroguanidine $h\,k\,l$ absent when $h + k$, $k + l$ and $h + l$ are odd

$0\,k\,l$ absent when $k + l$ is not divisible by 4

$h\,0\,l$ absent when $l + h$ is not divisible by 4

(b) p-nitrophenol $h\,0\,l$ absent when l is odd

$0\,k\,0$ absent when k is odd

(c) α-resorcinol $0\,k\,l$ absent when $k + l$ is odd.

$h\,0\,l$ absent when h is odd

$0\,0\,l$ absent when l is odd

(d) SnI_4 $0\,k\,l$ absent when k is odd

(e) $(CH_3\,C_6\,H_4\,NC)_4\,CoI_2$ $h\,k\,0$ absent when h is odd

$0\,0\,l$ absent when l is not divisible by 4.

what are their space groups?

7.   (i) If compound (h) of question 5 has the space group $P\overline{1}$, where do the Ni atoms lie in the unit cell? (For details of space group $P\overline{1}$ see Figs 4.11b and 4.12b).

(ii) A compound AB_2 has a cell content of 2 and has space group $P2_1\,2_1\,2$ (Figures 4.11(c) and 4.12(c)). Where must the A atoms lie in the cell?

(iii) A compound $XY_2\,Z_4$ has a cell content of 1 and has space group Pmm2 (Figures 4.11(d) and 4.12(d)) Where must the X and Y atoms lie in the cell?

(iv) Comment on the suggestion that a compound AB_2 with a cell content of 2 has the space group $Pna2_1$ (Figures 4.11(e) and 4.12(e)).

8. Which of the following molecules is expected to have a permanent electric dipole moment?

(a) chlorobenzene   (b) o-dichlorobenzene   (c) p-dichlorobenzene

(d) $COCl_2$ (e)H_2O_2 (f) $SOCl_2$

(g) $SO_2\,Cl_2$ (h) $CH_3\,Cl$ (i) $SF_5\,Cl$

(j) cis-$N_2\,F_2$ (k) trans-$N_2\,F_2$

9. Is it possible, on the basis of the existence or non-existence of a permanent dipole moment for the molecule or ion, to solve the following problems?

(i) Two configurations have been suggested for ClF_3, namely

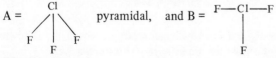

A = ⎨ pyramidal,   and B = F—Cl—F   planar

Which is correct?

(ii) A molecule whose formula is C_3H_4 is either cyclopropene or methyl-acetylene. Which is it?

(iii) H_3BO_3 is planar with the boron atom at the centre of an equilateral triangle of oxygen atoms. All the $B\hat{O}H$ angles are equal and all the B—O and O—H bond lengths are equal. Is the B—O—H system linear?

(iv) SF_4 has a trigonal bipyramidal structure in which either one axial or one equatorial position is not occupied by a fluorine atom. Which is correct?

(v) $[Co(ethylenediamine)_2 Cl_2]^+$ (green) goes to a red form on evaporation of a neutral aqueous solution at $90°C$. Can it be shown that the green form is the *trans*-isomer and the red form the *cis*-isomer?

(vi) Two structures have been suggested for the complex $[Ni(CN)_5]^{3-}$, namely, A, a square-based pyramid and B, a trigonal bipyramid. Which is correct?

10. Which of the following molecules or ions is expected to display optical activity?

(a) $[Co(C_2O_4)_3]^{3-}$
(b) Glycine, NH_2CH_2COOH
(c) alanine, CH_3CHNH_2COOH
(d) spiro—(4,4)—nonane
(e) spiro—(4,5)—decane
(f) *cis*-1, 2-dichlorocyclopropane
(g) *trans*-1, 2-dichlorocyclopropane
(h) cyclohexanone
(i) *cis*$[Co(NH_3)_4Cl_2]^+$
(j) *cis*-$[Co(C_2O_4)_2Cl_2]^{3-}$

11. Is it possible, on the basis of the existence or non-existence of optical activity of the molecule or ion, to solve the following problems?

(i) Problem 5 of question 9.
(ii) Problem 6 of question 9.

12. Suggest a complex of Pt which could be made in order to test, by studies of optical activity, whether the arrangements of four bonds about the Pt atom were tetragonal-pyramidal rather than square-planar.

13. Determine the distribution of vibrations among symmetry classes for the following molecules or ions, whose point group symmetry is given:

H_2O_2 (C_2) MoF_6 (O_h)
$COCl_2$ (C_{2v}) $AuCl_4^-$ (D_{4h})
BrO_3^- (C_{3v}) C_2H_6 (D_{3d})
p-dichlorobenzene (D_{2h}) CCl_4 (T_d)
$XeOF_4$ (C_{4v}) B_2Cl_4 (D_{2d})

14. Show, by constructing Γ_μ and Γ_α, which symmetry classes of the following point groups contain vibrations which are infra-red or Raman-active.

(a) C_{3v} (b) D_{2d} (c) D_2 (d) D_{6h}

15. Show, by forming the integrals $\int\phi_i M\phi_f$, $\int\phi_i\alpha\phi_f$, that the selection rules for vibrational activity under C_{2v} symmetry are

A_1	z	x^2, y^2, z^2
A_2		xy
B_1	x	xz
B_2	y	yz

16. Determine the selection rules for an electric dipole transition between ϕ_2 and ϕ_5, and between ϕ_2 and ϕ_6, in the system illustrated in Figure 6.14.

7

Symmetry and Theories of Bonding

The character table of a point group describes the irreducible representations of that group. As we have seen, it is set up by considering how any general point or function behaves on operation of the symmetry elements of the group. This is important because it means that, for a given point group, any function must behave (transform) either as one of the irreducible representations or as a combination of these representations. Tables 7.1 to 7.3 show how the functions x, y, z, xy, xz, yz, $x^2 - y^2$ and z^2 transform in the following point groups.

(a) C_{2v} where the z axis lies along the 2-fold axis, the yz plane is σ_v and the xz plane is σ'_v.

Table 7.1. Character table for C_{2v} symmetry

C_{2v}	I	C_2	$\sigma_v(yz)$	$\sigma'_v(xz)$		
A_1	1	1	1	1		$z, x^2, z^2, y^2, x^2-y^2$
A_2	1	1	-1	-1	R_z	xy
B_1	1	-1	-1	1	R_y	x, xz
B_2	1	-1	1	-1	R_x	y, yz

i.e. the functions z, z^2, $x^2 - y^2$ transform as the irreducible representation A_1; xy transforms as A_2; x and xz as B_1; and y and yz as B_2. The characters of functions such as xy and z^2 can be worked out as described in Chapter 5 because they are the products of the characters for the components of the function, i.e. $x.y$ and $z.z$.

(b) C_{3v} where the z axis lies along the 3-fold axis.

Table 7.2. Character table for C_{3v} symmetry

C_{3v}	I	$2C_3(z)$	$3\sigma_v$		
A_1	1	1	1		z, z^2
A_2	1	1	-1	R_z	
E	2	-1	0	(R_x, R_y)	$(xz, yz),(x^2 - y^2, xy)$

i.e. the functions z and z^2 transform as the irreducible representation A_1. The bracketed functions (x,y) (xz, yz) and $(x^2 - y^2, xy)$ transform together in pairs as the irreducible representation E. For example, operation of the C_3^1 element of the group on a point with co-ordinates (x, y) gives a new point with co-ordinates $(-\frac{1}{2}x + \frac{\sqrt{3}}{2}y, -\frac{\sqrt{3}}{2}x - \frac{1}{2}y)$. This shows that a function containing x is converted to a function containing both x and y and means that these must transform together as a degenerate pair in C_{3v} symmetry.

(c) O_h Table 7.3. Character table for O_h symmetry

O_h	I	$8C_3$	$3C_2$	$6C_2$	$6C_4$	i	$6S_4$	$8S_6$	$3\sigma_h$	$6\sigma_d$		
A_{1g}	1	1	1	1	1	1	1	1	1	1		s
A_{1u}	1	1	1	1	1	-1	-1	-1	-1	-1		
A_{2g}	1	1	1	-1	-1	1	-1	1	1	-1		
A_{2u}	1	1	1	-1	-1	-1	1	-1	-1	1		
E_g	2	-1	2	0	0	2	0	-1	2	0		$(z^2, x^2 - y^2)$
E_u	2	-1	2	0	0	-2	0	1	-2	0		
T_{1g}	3	0	-1	-1	1	3	1	0	-1	-1	$(R_x, R_y, R_z,)$	
T_{1u}	3	0	-1	-1	1	-3	-1	0	1	1		$(x, y, z,)$
T_{2g}	3	0	-1	1	-1	3	-1	0	-1	1		(xy, yz, xz)
T_{2u}	3	0	-1	1	-1	-3	1	0	1	-1		

The functions z^2 and $x^2 - y^2$ transform as a doubly degenerate pair with E_g symmetry. The functions x, y and z transform as a triply degenerate set of T_{1u} symmetry; likewise xy, yz and xz transform together as T_{2g}.

Atomic orbitals are described by wave functions which can be represented in polar co-ordinates as the product of a radial and an angular function, e.g. $p_x = R.\sin\theta.\cos\phi$ and, since $x = r.\sin\theta.\cos\phi$, $p_x = (R/r)x$ where R is a radial function dependent upon the distance r from the nucleus. R/r is invariant to all operations of any group and thus a p_x orbital transforms in the same way as the function x. Similarly p_y and p_z transform as the functions y and z respectively and $d_{xy}, d_{xz}, d_{yz}, d_{x^2-y^2}$ and d_{z^2} transform as the functions xy, xz, yz, $x^2 - y^2$ and z^2 respectively. Thus in the character tables for C_{2v}, C_{3v} and O_h we have picked out the irreducible representations which transform in the same way as the p- and d-orbitals of molecules having these symmetries. An s-orbital is spherically symmetrical and no operation of any group can alter that. s-orbitals must therefore transform in a group as the spherically symmetrical ($\equiv$ totally symmetric) irreducible representation which has the character +1 for all operations, i.e. A_1 for C_{2v} and C_{3v} and A_{1g} for O_h.

Most theories of bonding somehow make use of the knowledge of the way in which the atomic orbitals on the central atom of a molecule transform. In this section we shall consider the symmetry implications involved in crystal field theory, in the concepts of orbital mixing and in molecular orbital theory.

Symmetry and Crystal-field Theories

Crystal field theory considers the effect of electrostatic fields on the atomic orbitals of the central atom of a group. In a completely spherical field p-orbitals and d-orbitals would transform as a 3-fold degenerate set and a 5-fold degenerate set respectively. If the electrostatic field is not spherical the orbitals need not remain degenerate. If we have an atom in the centre of a field with octahedral, O_h, symmetry, for example, we have already seen that the d-orbitals are no longer degenerate (Table 7.3) but that they transform as a triply-degenerate set (symmetry T_{2g}) and a doubly degenerate set (E_g). Thus, by consideration of symmetry alone, we can state that an element (e.g. a transition metal ion) in an octahedral field will have its d-orbitals split into two sets of differing energy. What symmetry arguments cannot tell us is which of the two levels has the lower energy. It is only by considering the energetics of the repulsive forces on the d-orbitals that we can show that the T_{2g} orbitals have the lower energy. Reference to the character table for O_h symmetry shows us that the p-orbitals of an atom at the centre of an octahedral field will remain degenerate and would transform as the irreducible representation, T_{1u}. If we consider the p-orbitals of an atom in the centre of a field of C_{3v} symmetry, however, we find that (Table 7.2) they are no longer triply degenerate but that they must transform as a doubly degenerate E set (p_x and p_y) and a non degenerate A_1 set (p_z). Again this means that, from symmetry arguments alone, we can show that the p-orbitals of an atom in a field of C_{3v} symmetry are split into two levels, p_z and the degenerate pair p_x and p_y. Again, however, we cannot say from symmetry arguments alone whether the p_z orbital has the higher or lower energy and again reference must be made to the detailed energetics of the system to answer this question.

SYMMETRY AND ORBITAL MIXING

In theories of bonding which involve orbital mixing (hybridisation) the atomic orbitals of one or more of the atoms forming the molecule are considered to interact to form new types of orbitals. These concepts of orbital mixing are also based on considerations of symmetry, as can be seen from the following examples.

If we consider that the p_z-orbital of the molecule BF_3 lies along the 3-fold axis and that the p_x- and p_y-orbitals lie along the directions shown it is obvious

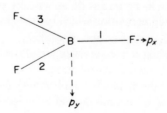

that the atomic orbitals of the B atom are not in the correct positions to form σ bonds to the three fluorine atoms. What we must do is to obtain the total character of the orbitals involved in the bonds to the fluorine atoms and then, by reference to the character table for D_{3h} symmetry, work out which combination of atomic orbitals of the B atom is forming the bonds.

The symmetry elements present in D_{3h} symmetry are I, $2C_3$, $3C_2$, $2S_3$, σ_h and $3\sigma_v$. We can determine the total character of the bonding orbitals by labelling the three bonds 1, 2, 3 and finding out how they behave on operation of all of the elements present.

If we operate the identity on the three bonds they remain in the same positions and this is also true for operation of σ_h. Expressing this in matrix form we get a matrix of character 3.

$$I(1, 2, 3) = (1, 2, 3) \begin{bmatrix} 1 & 0 & 0 \\ 0 & 1 & 0 \\ 0 & 0 & 1 \end{bmatrix} = \sigma_h(1, 2, 3)$$

On operation of the 3-fold axis C_3^1, bond 1 becomes bond 2, bond 2 becomes bond 3 and bond 3 becomes bond 1.

$$C_3^1(1, 2, 3) = (1, 2, 3) \begin{bmatrix} 0 & 0 & 1 \\ 1 & 0 & 0 \\ 0 & 1 & 0 \end{bmatrix}$$

This matrix has character 0 and similarly the matrices representing the operation of C_3^2 and the S_3 axes have character 0.

Operation of the 2-fold axis or the vertical mirror plane through bond 1 leaves it unchanged but does interconvert bonds 2 and 3, i.e.

$$\sigma_v(1, 2, 3) = (1, 2, 3) \begin{bmatrix} 1 & 0 & 0 \\ 0 & 0 & 1 \\ 0 & 1 & 0 \end{bmatrix} = C_2(1, 2, 3)$$

This matrix has character 1 as do the matrices representing the operations of the other 2-fold axes and vertical mirror planes.

The total character of the orbitals forming the three σ bonds in BF_3 is thus:

	I	$2C_3$	$3C_2$	$2S_3$	σ_h	$3\sigma_v$
χ	3	0	1	0	3	1

Reference to the character table for D_{3h} symmetry (Table 7.4) shows that χ can only be produced by the sum of the irreducible representations $A'_1 + E'$. Thus the total character of the three bond orbitals is made up by contributions from the s-orbital (A'_1) and the doubly degenerate p_x- and p_y-orbitals (E'), i.e. the boron atom is using three sp^2 hybrid orbitals.

Table 7.4. Character table for D_{3h} symmetry

D_{3h}	I	$2C_3$	$3C_2$	$2S_3$	σ_h	$3\sigma_v$		
A'_1	1	1	1	1	1	1		s, x^2+y^2, z^2
A'_2	1	1	-1	1	1	-1	R_z	
A''_1	1	1	1	-1	-1	-1		
A''_2	1	1	-1	-1	-1	1		z
E'	2	-1	0	-1	2	0		$(x, y)\ (x^2-y^2, xy)$
E''	2	-1	0	1	-2	0	(R_x, R_y)	(xz, yz)

In the previous chapter we saw that there was no need to work out the total character of a vibration from first principles. Instead we could obtain the character for each operation by counting the number of atoms which were not shifted by the operation of a symmetry element. In the same way we can obtain the total character of a system of bond orbitals by counting the number of orbitals which do not change on operation of the symmetry element.

For example for BF_3

(i) operation of I or of σ_h leaves all three bonds in the same position

$\therefore$ $$\chi(I) = 3 = \chi(\sigma_h)$$

(ii) operation of C_3 or of S_3 changes the relative positions of all the orbitals

$\therefore$ $$\chi(C_3) = 0 = \chi(S_3);$$

(iii) operation of C_2 or of σ_v leaves one orbital in its original position

$\therefore$ $$\chi(C_2) = 1 = \chi(\sigma_v).$$

We thus obtain the same result for the total character

	I	$2C_3$	$3C_2$	$2S_3$	σ_h	$3\sigma_v$
χ_{total}	3	0	1	0	3	1

as we did from first principle arguments.

By similar methods we can obtain the total character of the four orbitals in methane.

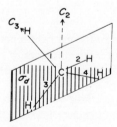

The elements present in the point group T_d are

$$I \quad 8C_3 \quad 6\sigma_d \quad 6S_4 \quad 3C_2$$

and we can show that:
(i) all four orbitals remain the same on operation of I;
(ii) only one orbital remains the same on operation of any of the 3-fold axes (e.g. orbital 1 for the C_3 shown);
(iii) two of the orbitals remain the same on operation of a σ_d (e.g. orbitals 3 and 4 for the σ_d shown);
(iv) none of the orbitals remains the same on operation of S_4 or C_2.

The total character of the bond orbitals is thus:

	I	$8C_3$	$6\sigma_d$	$6S_4$	$3C_2$
χ_{total}	4	1	2	0	0

and reference to the character table for T_d symmetry shows that the total character must be made up of contributions from A_1 (s-orbital) and T_2 (triply degenerate p_x, p_y and p_z orbitals).

Table 7.5. Character table for T_d symmetry

T_d	I	$8C_3$	$6\sigma_d$	$6S_4$	$3C_2$		
A_1	1	1	1	1	1		s
A_2	1	1	−1	−1	1		
E	2	−1	0	0	2		$(z^2, x^2 - y^2)$
T_1	3	0	−1	1	−1	(R_x, R_y, R_z)	
T_2	3	0	1	−1	−1		$(x,y,z)\ (xy,\ xz,\ yz)$

Note that, in general for tetrahedral molecules XY_4 $\chi_{total} = A_1 + T_2$ can be satisfied by the combinations

$$s(A_1) + (p_x, p_y, p_z)(T_2) \text{ or } s(A_1) + (d_{xy}, d_{xz}, d_{yz})(T_2)$$

For methane only the former combination is possible because the carbon atom has no energetically accessible d-orbitals but for molecules such as SiF_4, $GeCl_4$, etc. we must decide on energetic grounds whether there is any d-orbital participation in the σ bonding.

The character of any function in a group can always be obtained by finding out how it varies with each operation of the group. We have seen that, if a single function, atom or orbital is invariant to an operation, it contributes +1 to the character. If, however, the operation changes the sign of the function but otherwise leaves it unchanged then a contribution of −1 is made to the character. We can illustrate these effects by working out the character of the p_z-orbital in BF_3. Figure 7.1 shows that the orbital does not change position or sign on operation

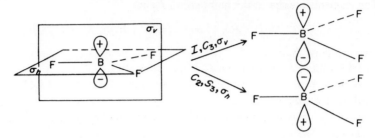

Figure 7.1　　　Operation of the elements of D_{3h} symmetry in the p_z orbital of BF_3.

of I, C_3 or σ_v, but that it changes its sign but not its position on operation of C_2, S_3 or σ_h. This gives us the total character of the p_z-orbital as:

	I	$2C_3$	$3C_2$	$2S_3$	σ_h	$3\sigma_v$
χ_{p_z}	1	1	−1	−1	−1	1

which is the irreducible representation A_2'' of the point group D_{3h} (see Table 7.4).

Symmetry and Molecular Orbital Theory

The molecular orbital approach to bond descriptions is based on the assumption that orbitals of the same symmetry in two adjacent atoms or groups can overlap to form bonding orbitals. The bonding orbital will be a linear combination of the atomic or group orbitals and associated with this there will be a corresponding linear combination with antibonding properties. From the point of view of symmetry arguments alone, it is not necessary to know the value of the normalising constants in the linear combinations, and for this reason we have chosen to omit them in most of the discussion in this book. The constants are, however, included in figures or tables for some of the examples discussed.

The simplest molecular orbital descriptions are those for the homonuclear diatomic molecules such as N_2 and O_2. These descriptions involve only the overlap of the s- and p- orbitals of the atoms forming the bond as illustrated in Figure 7.2. The symmetry symbols given for the molecular orbitals in this diagram are those of the corresponding irreducible representations of the point group $D_{\infty h}$ (Table 7.6). The energy level diagram of Figure 7.2 is simplified and, even for some homonuclear diatomic molecules it is necessary to allow some $s-p$ mixing. Mixing of s- and p_z-atomic orbitals on both the C and O atoms in the heteronuclear diatomic molecule CO is necessary in order to explain the energy of the C–O bond and the fact that the molecule acts as a donor through the carbon atom. Figure 7.3 shows that s- and p_z-orbitals on any atom can be mixed to give two new orbitals $s + p_z$ and $s - p_z$. These new orbitals can then

Overlap of s- and p-orbitals in diatomic molecules

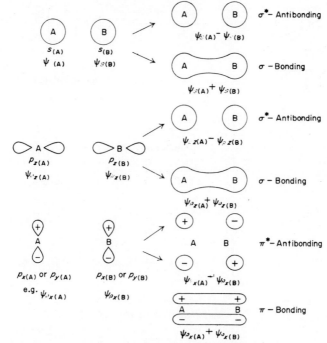

Molecular orbital description of homonuclear diatomic molecules

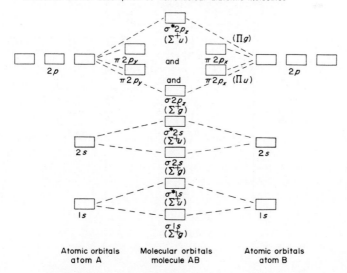

Figure 7.2   Overlap of *s* and *p* atomic orbitals and the molecular orbital description of homonuclear diatomic molecules

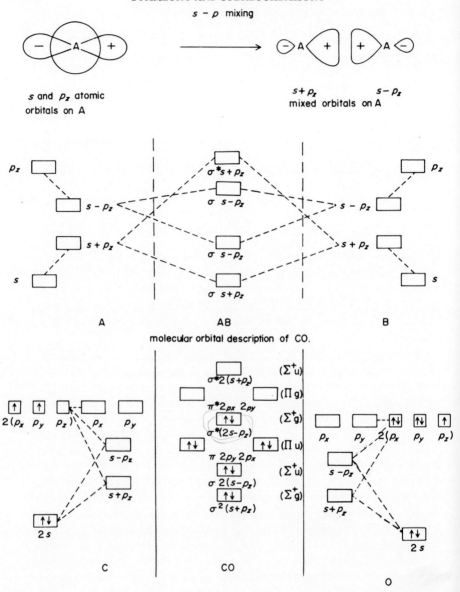

Figure 7.3   *s-p* mixing and the molecular orbital description of CO

Table 7.6. Character table for $D_{\infty h}$

$D_{\infty h}$	E	$2C_\infty{}^\Phi$	...	$\infty\sigma_v$	i	$2S_\infty{}^\Phi$	...	∞C_2		
$\Sigma_g{}^+$	1	1	...	1	1	1	...	1		x^2-y^2, z^2
$\Sigma_g{}^-$	1	1	...	-1	1	1	...	-1	R_z	
Π_g	2	$2\cos\Phi$	...	0	2	$-2\cos\Phi$	...	-1	(R_x, R_y)	(xz, yz)
Δ_g	2	$2\cos 2\Phi$	...	0	2	$2\cos 2\Phi$	...	0		(x^2-y^2, xy)
...	...	...	...	...	...	...	...	...		
$\Sigma_u{}^+$	1	1	...	1	-1	-1	...	-1	z	
$\Sigma_u{}^-$	1	1	...	-1	-1	-1	...	-1		
Π_u	2	$2\cos\Phi$	...	0	-2	$2\cos\Phi$	...	0	(x, y)	
Δ_u	2	$2\cos 2\Phi$	...	0	-2	$-2\cos 2\Phi$	...	0		
...	...	...	...	...	...	...	...	...		

combine with orbitals of identical symmetry on another atom to give σ bonding $[\sigma(s+p_z), \sigma(s-p_z)]$ and σ antibonding orbitals $[\sigma^*(s+p_z), \sigma^*(s-p_z)]$. Figure 7.3 also shows the energy level diagram for CO. In this diagram the symmetry symbols shown are those of the irreducible representations of $C_{\infty v}$. It should be stressed once again that symmetry arguments alone can only predict the types of molecular orbital which should be found and that they give no information on the energies of these orbitals. The order of molecular orbitals in the energy diagrams of Figures 7.2 and 7.3 are based on information in addition to that obtained from symmetry considerations.

In the molecular orbital description of polyatomic molecules we have to devise a means of working out which orbitals on the various atoms or groups have the same symmetry. The central atom of a polyatomic system is unique in that it lies on all of the symmetry elements of the point group of the system. This means that the atomic orbitals of the central atom will transform as irreducible representations of that point group. For example, the oxygen atom of the water molecule lies on all of the symmetry elements of the point group C_{2v}. If we adopt the axes used to construct the character table for C_{2v} (Table 7.1) then the s- and p_z-orbitals of oxygen transform as A_1, p_x as B_1, and p_y as B_2. The other atoms of a polyatomic system (e.g. the hydrogen atoms of water) do not lie on all of the symmetry elements of the group and cannot transform as one of its irreducible representations. We can, however, consider the various linear combinations of the orbitals on other atoms and find out whether any of these transform as irreducible representations of the group and whether any of the combinations (*Group Orbitals*) have the same symmetries as the atomic orbitals of the central atom. Only two linear combinations of the s-orbitals on the hydrogen atoms of water are possible $s_{(A)}+s_{(B)}$ and $s_{(A)}-s_{(B)}$ (see Figure 7.4). The wave function

$$\Psi_1 = \frac{1}{\sqrt{2}}(\psi_{s_{(A)}} + \psi_{s_{(B)}})$$

has the same symmetry as the s- or p_z-orbitals of the oxygen atom, i.e. A_1.

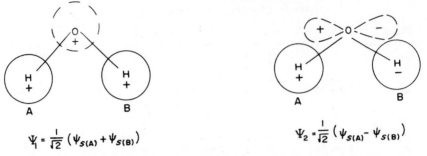

$$\Psi_1 = \frac{1}{\sqrt{2}} \left(\Psi_{s(A)} + \Psi_{s(B)} \right)$$

$$\Psi_2 = \frac{1}{\sqrt{2}} \left(\Psi_{s(A)} - \Psi_{s(B)} \right)$$

Figure 7.4   Linear combinations of hydrogen s orbitals in the water molecule

The wave function

$$\Psi_2 = \frac{1}{\sqrt{2}} \left(\psi s_{(A)} - \psi s_{(B)} \right)$$

has the same symmetry as the p_y-orbital (B_2). These statements can be checked by inspection of the diagrams in Figure 7.4 but can also be verified by operation of the symmetry elements in the group on the linear combinations. If we consider the linear combination $s_{(A)} + s_{(B)}$ (+ + in the diagram) and operate the symmetry elements of the group we get the same pattern each time, i.e. + +. Likewise, operation of the elements I and $\sigma_v(yz)$ on the combination $s_{(A)} - s_{(B)}$($+ -$ in the diagram) gives the same pattern ($+ -$) but operation of C_2 and $\sigma_v(xz)$ gives the pattern ($- +$) in which the s- orbitals are in the same relative positions but have different signs.

The characters of these group orbitals are therefore:

	I	C_2	$\sigma_v(yz)$	$\sigma_v'(xz)$	
$s_{(A)} + s_{(B)}$	1	1	1	1	A_1
$s_{(A)} - s_{(B)}$	1	-1	1	-1	B_2

Once we know which group orbitals have the same symmetry as the atomic orbitals we can construct a molecular orbital diagram. For example, in the water molecule the group orbital $s_{(A)} - s_{(B)}$ has the same symmetry as the p_y-orbital and so we have one bonding and one anti-bonding molecular orbital with symmetry B_2. Similarly we can have bonding and anti-bonding molecular orbitals with symmetry A_1 but we cannot, by symmetry arguments alone, decide whether the oxygen atomic orbital used in bond formation is the s- or p_z-orbital, both of which transform as A_1. If we assume that the s-orbital is used in bond orbital formation then the p_z-orbital must be a non-bonding orbital of symmetry A_1. Similarly, since there is no group orbital combination which transforms as B_1, the p_x-orbital of oxygen is also non-bonding. Figure 7.5 illustrates the molecular orbital energy diagram for water.

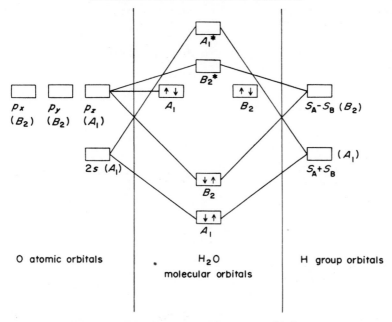

Figure 7.5   Molecular orbital description of the water molecule

The molecule SO_2 also has C_{2v} symmetry and we can extend the arguments used in the molecular orbital description of water to deal with cases like SO_2 for which a large choice of bonding atomic orbitals is possible. The orbitals on the oxygen atoms which are responsible for forming the σ bonds to the sulphur are presumably either p-orbitals or sp hybrids of some sort. It is not necessary, from the point of view of symmetry, to know what type of oxygen orbital is forming the bond. It is sufficient to know that on each oxygen atom there is a σ bond-forming orbital. We can see from Figure 7.6 that the combinations of these σ-type orbitals which have the same symmetry as orbitals in the sulphur atom are $\sigma_{(A)} + \sigma_{(B)}$ $(A_1 : s, p_z, d_z^2, d_x^2 - y^2)$ and $\sigma_{(A)} - \sigma_{(B)} (B_2 : p_y, d_{yz})$. The symmetries of the σ-bonding orbitals are thus the same as those for the water molecule, as in fact they must be. In the case of SO_2, however, there is the additional energetic problem of possible d-orbital participation in the bonding. There are no combinations of σ-type oxygen orbitals that have the same symmetry as the p_x- or d_{xz}-orbitals on the S atom. The oxygen atoms do, however, have other orbitals available which are capable of forming π bonds and the combination $\pi_{(A)} + \pi_{(B)}$ (i.e. p_x-orbitals or suitable hybrid orbitals on the oxygen atoms) has the symmetry B_1 (p_x or d_{xz} on sulphur). The example shown in Figure 7.6 is of the symmetry equivalence of $\pi_{(A)} + \pi_{(B)}$ and p_x. There are no combinations of either σ or π type orbitals in the oxygen atoms which have the same symmetry (A_2) as the d_{xy} in sulphur. The A_2 orbital must therefore remain non-bonding in the molecular orbital

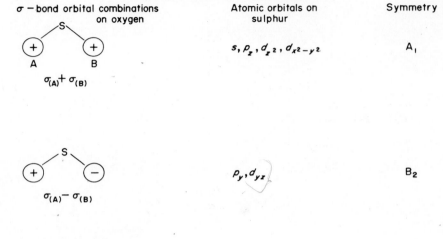

σ−bond orbital combinations on oxygen	Atomic orbitals on sulphur	Symmetry
	$s, p_z, d_{z^2}, d_{x^2-y^2}$	A_1
	p_y, d_{yz}	B_2

π−bond orbital combinations on oxygen

	p_x, d_{xz}	B_1

Figure 7.6   Group orbitals in the molecular orbital description of SO_2.

diagram. The bonding molecular orbitals in the molecular orbital description must all have one of the symmetries A_1, B_1 and B_2 and, if we neglect d-orbital participation in the bonding, the detailed molecular orbital energy level diagram for SO_2 is that shown in Figure 7.7.

The reader can verify that for a square planar molecule XY_4 the group orbitals in Table 7.7 have the same symmetries as atomic orbitals on the central atom. (The character table for D_{4h} is given as Table 7.8).

LIGAND-FIELD MOLECULAR ORBITAL THEORY

The ligand-field molecular orbital theory is simply an extension of molecular orbital ideas to explain the properties of transition metal molecules and ions. The group orbitals in this case are the combinations of suitable orbitals in the ligands. For example, every ligand in an octahedral transition-metal complex contains a lone pair orbital which is capable of forming a σ bond with the transition metal. The ligand group orbitals are found by taking linear combinations of

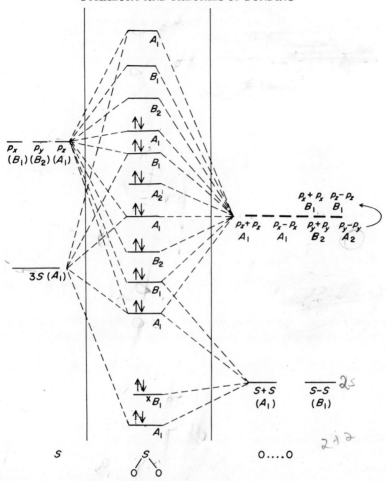

Figure 7.7   Molecular orbital energy level diagram for SO_2 showing the allocation of the eighteen available electrons to molecular orbitals

Table 7.7. Group orbitals for square-planar molecules XY_4

σ group orbitals	π group orbitals	Atomic orbital on central atom	Symmetry
$\sigma_{(A)} + \sigma_{(B)} + \sigma_{(C)} + \sigma_{(D)}$		s, d_{z^2}	A_{1g}
$\sigma_{(A)} + \sigma_{(B)} - \sigma_{(C)} - \sigma_{(D)}$		$d_{x^2-y^2}$	B_{1g}
	$\pi_A - \pi_B + \pi_C - \pi_D$	d_{xy}	B_{2g}
$\sigma_{(A)} - \sigma_{(C)}$	$\pi_A - \pi_C$	$\left.\begin{array}{l} p_x \\ p_y \end{array}\right\}$	E_u
$\sigma_{(B)} - \sigma_{(D)}$	$\pi_B - \pi_D$		
–	–	$\left.\begin{array}{l} d_{xy} \\ d_{yz} \end{array}\right\}$	E_g
–	–	p_z	A_{2u}

Table 7.8. Character table for D_{4h} symmetry

D_{4h}	I	$2C_4$	C_2	$2C_2'$	$2C_2''$	i	$2S_4$	σ_h	$2\sigma_v$	$2\sigma_d$		
A_{1g}	1	1	1	1	1	1	1	1	1	1		x^2+y^2, z^2
A_{2g}	1	1	1	-1	-1	1	1	1	-1	-1	R_z	
B_{1g}	1	-1	1	1	-1	1	-1	1	1	-1		x^2-y^2
B_{2g}	1	-1	1	-1	1	1	-1	1	-1	1		xy
E_g	2	0	-2	0	0	2	0	-2	0	0	(R_x, R_y)	(xz, yz)
A_{1u}	1	1	1	1	1	-1	-1	-1	-1	-1		
A_{2u}	1	1	1	-1	-1	-1	-1	-1	1	1		z
B_{1u}	1	-1	1	1	-1	-1	1	-1	-1	1		
B_{2u}	1	-1	1	-1	1	-1	1	-1	1	-1		
E_u	2	0	-2	0	0	-2	0	2	0	0		(x, y)

the wave functions of these lone-pair orbitals neglecting as a first approximation any ligand–ligand interaction. If we consider the octahedral complex:

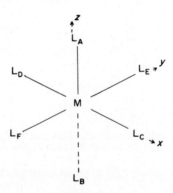

in which the wave function representing the lone-pair orbital on ligand $A(L_A)$ is $\sigma_{(A)}$ etc. then we can tabulate (Table 7.9) the ligand group orbitals which have the same symmetry as the atomic orbitals in M. The ligand group orbitals are also shown diagrammatically in Figure 7.8.

We can thus form σ molecular orbitals by combining the atomic orbitals s, $p_x, p_y, p_z, d_{x^2-y^2}$, and d_{z^2} with the ligand group orbitals of corresponding symmetry. This gives rise to the familiar ligand-field molecular orbital diagram

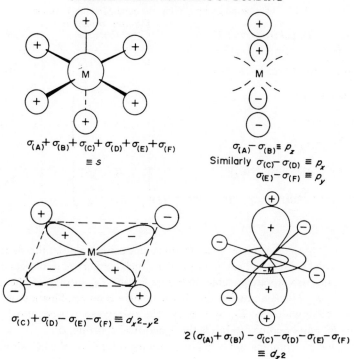

$\sigma_{(A)} + \sigma_{(B)} + \sigma_{(C)} + \sigma_{(D)} + \sigma_{(E)} + \sigma_{(F)}$

$\equiv s$

$\sigma_{(A)} - \sigma_{(B)} \equiv p_z$

Similarly $\sigma_{(C)} - \sigma_{(D)} \equiv p_x$

$\sigma_{(E)} - \sigma_{(F)} \equiv p_y$

$\sigma_{(C)} + \sigma_{(D)} - \sigma_{(E)} - \sigma_{(F)} \equiv d_{x^2 - y^2}$

$2(\sigma_{(A)} + \sigma_{(B)}) - \sigma_{(C)} - \sigma_{(D)} - \sigma_{(E)} - \sigma_{(F)}$

$\equiv d_{z^2}$

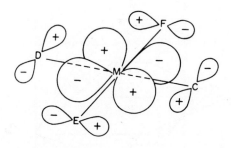

$\pi_{p(C)} - \pi_{p(D)} + \pi_{p(E)} - \pi_{p(F)} \equiv d_{xy}$

Similar combinations give d_{xz} and d_{yz}.

Figure 7.8   Interaction between the atomic orbitals of an atom and ligand group orbitals for O_h symmetry

SYMMETRY AND STEREOCHEMISTRY

Table 7.9. σ-ligand group orbitals for octahedral (O_h) symmetry

Ligand group orbital	Atomic orbital of central atom	Symmetry
$\frac{1}{\sqrt{6}}(\sigma_{(A)} + \sigma_{(B)} + \sigma_{(C)} + \sigma_{(D)} + \sigma_{(E)} + \sigma_{(F)})$	s	A_{1g}
$\frac{1}{\sqrt{2}}(\sigma_{(A)} - \sigma_{(B)})$	p_z	
$\frac{1}{\sqrt{2}}(\sigma_{(C)} - \sigma_{(D)})$	p_x	T_{1u}
$\frac{1}{\sqrt{2}}(\sigma_{(E)} - \sigma_{(F)})$	p_y	
$\frac{1}{2}(\sigma_{(C)} - \sigma_{(F)} + \sigma_{(D)} - \sigma_{(E)})$	$d_{x^2 - y^2}$	E_g
$\frac{1}{\sqrt{12}}(2\sigma_{(A)} + 2\sigma_{(B)} - \sigma_{(C)} - \sigma_{(D)} - \sigma_{(E)} - \sigma_{(F)})$	d_{z^2}	

of Figure 7.9. Again, symmetry arguments alone cannot predict the energies of
the molecular orbitals but can only predict which combinations of ligand and
metal atomic orbitals are possible. In Figure 7.9 the T_{2g} orbitals on the metal
(d_{xy}, d_{xz}, d_{yz}) remain non-bonding because there is no combination of ligand
σ-type orbitals which has T_{2g} symmetry. The energy gap between this T_{2g} level
and the E_g^* anti-bonding orbitals is the ligand-field splitting, Δ.

Although no combination of σ bond-forming orbitals on the ligands has the

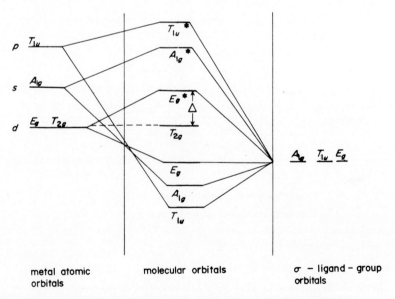

Figure 7.9   The σ-ligand-field-molecular-orbital diagram for O_h symmetry

same symmetry as the d_{xy}-, d_{xz}- and d_{yz}-orbitals, the following combinations of π bond forming orbitals do have the correct symmetry (see Figure 7.8):

$$\pi_{p(A)} - \pi_{p(B)} + \pi_{p(C)} - \pi_{p(D)} \equiv d_{xz}$$

$$\pi_{p(A)} - \pi_{p(B)} + \pi_{p(E)} - \pi_{p(F)} \equiv d_{yz}$$

$$\pi_{p(E)} - \pi_{p(F)} + \pi_{p(C)} - \pi_{p(D)} \equiv d_{xy}$$

We can now complete the ligand-field molecular orbital diagram for O_h symmetry by allowing for the formation of T_{2g} π bonding and π^* antibonding orbitals (Figure 7.10).

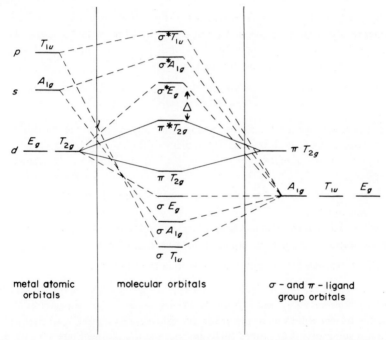

Figure 7.10   The ligand-field-molecular-orbital diagram for O_h symmetry including π-bonding

The ligand-field molecular orbital description of compounds of other symmetries can be set up in a similar manner and that for square-planar molecules and ions can be deduced from Table 7.7.

Symmetry and Molecular Orbital Calculations

The calculation of energy levels and of molecular orbital coefficients by Hückel and related methods gives a determinantal equation whose order is equal

to the number of molecular orbitals formed. By taking the symmetry properties
of the molecule into account, the determinantal equation can be broken down
into a number of equations of lower order, making the computation very much
simpler.

One of the most important applications of this treatment is concerned with
the solution of the determinantal equation for conjugated systems. The problem
here is to determine the energies of the molecular orbitals formed from those
atomic p-orbitals of the carbon skeleton which are not involved in the formation
of σ bonds. The procedure for solving this problem with the aid of the symmetry
properties of the molecule can be broken down into four steps.

1. Find the point group of the molecule under consideration and label the
atomic p-orbitals ϕ_1, ϕ_2 ---, ϕ_k, ---ϕ_n.

2. By applying the symmetry operations of the sub-group consisting of the
proper rotations of the molecular point group, find the character $\chi(p)$ of the
reducible representation formed by the atomic p-orbitals, and break this down
into its component irreducible representations using

$$a_j = \frac{1}{h} \sum_R g_R \chi(R) \chi_j(R)$$

The reason why we use the rotation sub-group only is because all p-orbitals have
the same symmetry with respect to the plane of the molecule, and so the
inclusion of this symmetry element does not help to reduce the determinantal
equation.

3. Construct the required number of 'symmetry-adapted orbitals' indicated by
the result of (2).

4. Set up the determinantal equation for the orbitals of each irreducible
representation and solve the equation for the energy levels.

As a first example let us apply this technique to butadiene.

Step 1.   There are two possible molecular point groups, corresponding to a *'cis'*
form of symmetry C_{2v} and a *'trans'* form of symmetry C_{2h} respectively. In either
case the proper rotations of the group consist solely of I and C_2, so that the
rotation sub-group is C_2. Since there are four carbon atoms, there are four atomic
p-orbitals which we will label Φ_1, Φ_2, Φ_3, Φ_4.

Step 2.   The character of the reducible representation for the p-orbitals is simply
the number of orbitals remaining invariant to each symmetry operation. With the
carbon atoms designated as:

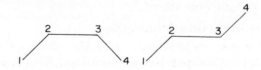

we see that the behaviour of the orbitals under the two symmetry operations is as follows:

	I	C_2
Φ_1	Φ_1	Φ_4
Φ_2	Φ_2	Φ_3
Φ_3	Φ_3	Φ_2
Φ_4	Φ_4	Φ_1
Number of invariant orbitals	4	0

Using the character table of the group C_2 (Table 7.10) we find that the reducible representation formed by the orbitals Φ_1, Φ_2, Φ_3, Φ_4 contains the irreducible representations A and B twice each.

Table 7.10. Character table of the point group C_2

C_2	I	C_2		
A	1	1	R_z	z, x^2, y^2, z^2, xy
B	1	-1	R_x, R_y	x, y, yz, xz

$$a_{(A)} = \tfrac{1}{2}[1.1.4 + 1.1.0] = 2$$
$$a_{(B)} = \tfrac{1}{2}[1.1.4 - 1.1.0] = 2$$

We therefore need to construct two symmetry-adapted orbitals of class A and two of class B.

Step 3. The most commonly used procedure for constructing such orbitals may be expressed by the equation

$$\phi_j = \sum_R R\Phi_G \cdot \chi_j(R) \tag{7.1}$$

Let the orbital to be constructed be ϕ_j, which belongs to the jth irreducible representation. Suppose we choose a member Φ_G of the original set and generate ϕ_j from Φ_G, then the procedure is to write down for every operation of the point group, the result $(R\Phi_G)$ of performing the operation R on Φ_G, and then to multiply this result by $\chi_j(R)$, the character of R in the jth irreducible representation. ϕ_j is then the sum of all such terms.

Let us generate an orbital $\phi_1(A)$ starting from Φ_1 as generator. There are only two operations of the group C_2, these being I, C_2; we can thus write equation (7.1) explicitly as:

$$\phi_1(A) = I\Phi_1 \cdot \chi_A(I) + C_2\Phi_1 \cdot \chi_A(C_2) \tag{7.2}$$

Now the result of performing the operation I on Φ_1 is Φ_1, while C_2 performed on Φ_1 gives Φ_4; thus we can re-write (7.2) as:

$$\phi_1(A) = \Phi_1\chi_A(I) + \Phi_4\chi_A(C_2) \tag{7.3}$$

The character table for the point group C_2 shows us that the character of I in the A representation is 1, and so is the character of C_2 in the A representation. Substituting these values as (7.3) we have

$$\begin{aligned}\phi_1(A) &= \Phi_1 \times 1 + \Phi_4 \times 1 \\ &= \Phi_1 + \Phi_4\end{aligned} \tag{7.4}$$

This is not normalised; in order to normalise ϕ_1 the sums of the squares of the coefficients of the Φ_k must be 1; thus on normalisation

$$\phi_1(A) = \frac{1}{\sqrt{2}}(\Phi_1 + \Phi_4) \tag{7.5}$$

We need to generate a second orbital of class A, $\phi_2(A)$, and this must be orthogonal to $\phi_1(A)$. For two orbitals $\phi_i = \sum_{k}c_{ik}\Phi_k$, $\phi_j = \sum_{k}c_{jk}\Phi_k$, to be orthogonal, the coefficients c_{ik}, c_{jk} must satisfy the condition

$$\sum_{k}c_{ik}c_{jk} = 0$$

Repetition of the procedure in this step, starting with $\Phi_G = \Phi_2$ gives:

$$\phi_2(A) = \frac{1}{\sqrt{2}}(\Phi_2 + \Phi_3).$$

It is clear that ϕ_2 is orthogonal to ϕ_1 since only Φ_1 and Φ_4 appear in ϕ_1 while only Φ_2 and Φ_3 appear in ϕ_2. We now need to construct two orbitals ϕ_3, ϕ_4 of class B; since the character of C_2 in the B representation is -1, the use of Φ_1 and Φ_2 as generators of ϕ_3 and ϕ_4 will produce

$$\phi_3(B) = \frac{1}{\sqrt{2}}(\Phi_1 - \Phi_4)$$

$$\phi_4(B) = \frac{1}{\sqrt{2}}(\Phi_2 - \Phi_3)$$

Step 4. The construction and solution of the determinants may now be carried out. For the butadiene molecule we could write the determinantal equation in full as

$$\begin{vmatrix} H_{11} - ES_{11} & H_{12} - ES_{12} & H_{13} - ES_{13} & H_{14} - ES_{14} \\ H_{21} - ES_{21} & H_{22} - ES_{22} & H_{23} - ES_{23} & H_{24} - ES_{24} \\ H_{31} - ES_{31} & H_{32} - ES_{32} & H_{33} - ES_{33} & H_{34} - ES_{34} \\ H_{41} - ES_{41} & H_{42} - ES_{42} & H_{43} - ES_{43} & H_{44} - ES_{44} \end{vmatrix} = 0$$

where

$$H_{ii} = \int \phi_i \mathcal{H} \phi_i d\tau$$

$$H_{ik} = \int \phi_i \mathcal{H} \phi_k d\tau = \int \phi_k \mathcal{H} \phi_i d\tau$$

$$S_{ii} = \int \phi_i^2 d\tau$$

$$S_{ik} = \int \phi_i \phi_k d\tau$$

Now the advantage in factoring out this 4×4 determinant into two 2×2 determinants is that cross terms H_{ik} where ϕ_i and ϕ_k belong to different irreducible representations (symmetry classes) are identically zero. Further, if the set of wave functions $\phi_i, \phi_j \ldots$ belonging to any one class are orthogonal and normalised, $S_{ii} = 1$ and S_{ik} $(i \neq k) = 0$.

Thus in butadiene, where we now have four symmetry-adapted orbitals ϕ_1, ϕ_2 of class A and ϕ_3, ϕ_4 of class B, the terms $H_{13}, H_{14}, H_{23}, H_{24}, H_{31}, H_{41}, H_{32}, H_{42}$ are all zero. If we also put in the values of S_{ii} and S_{ik} appropriate to the orthonormal sets, the determinant becomes

$$\begin{vmatrix} H_{11} - E & H_{12} & 0 & 0 \\ H_{21} & H_{22} - E & 0 & 0 \\ 0 & 0 & H_{33} - E & H_{34} \\ 0 & 0 & H_{43} & H_{44} - E \end{vmatrix} = 0$$

and the two 2×2 blocks can be solved separately.

From the definition above, we have

$$H_{11} = \int \phi_1 \mathcal{H} \phi_1 d\tau = \int \frac{1}{\sqrt{2}} [\Phi_1 + \Phi_4] \mathcal{H} \frac{1}{\sqrt{2}} [\Phi_1 + \Phi_4] \, d\tau$$

$$= \frac{1}{2} \int (\Phi_1 \mathcal{H} \Phi_1 + \Phi_1 \mathcal{H} \Phi_4 + \Phi_4 \mathcal{H} \Phi_1 + \Phi_4 \mathcal{H} \Phi_4) d\tau$$

If we now make the definitions

$$\int \Phi_i \mathcal{H} \Phi_i d\tau = \alpha_i$$

$$\int \Phi_i \mathcal{H} \Phi_k d\tau = \beta_{ik}$$

and introduce the Huckel assumptions that all α_i are equal and that all β_{ik} (i adjacent to k) are equal, while all β_{ik} (i non-adjacent to k) are zero, we see that $H_{11} = \frac{1}{2}(\alpha + 0 + 0 + \alpha) = \alpha$.

Evaluating the remaining terms

$$H_{12} = H_{21} = \int \phi_1 \mathcal{H} \phi_2 \, d\tau = \int \frac{1}{\sqrt{2}} [\Phi_1 + \Phi_4] \mathcal{H} \frac{1}{\sqrt{2}} [\Phi_2 + \Phi_3] \, d\tau$$

$$= \tfrac{1}{2} \int (\Phi_1 \mathcal{H} \Phi_2 + \Phi_1 \mathcal{H} \Phi_3 + \Phi_4 \mathcal{H} \Phi_2 + \Phi_4 \mathcal{H} \Phi_3) \, d\tau$$

$$= \tfrac{1}{2} [\beta + 0 + 0 + \beta]$$

$$= \beta$$

$$H_{22} = \int \Phi_2 \mathcal{H} \Phi_2 \, d\tau = \int \frac{1}{\sqrt{2}} [\Phi_2 + \Phi_3] \mathcal{H} \frac{1}{\sqrt{2}} [\Phi_2 + \Phi_3] \, d\tau$$

$$= \tfrac{1}{2} \int (\Phi_2 \mathcal{H} \Phi_2 + \Phi_3 \mathcal{H} \Phi_2 + \Phi_2 \mathcal{H} \Phi_3 + \Phi_3 \mathcal{H} \Phi_3) \, d\tau$$

$$= \tfrac{1}{2} [\alpha + \beta + \beta + \alpha]$$

$$= \alpha + \beta$$

The reader may verify that $H_{33} = H_{11} = \alpha$

$$H_{34} = H_{43} = H_{12} = H_{21} = \beta$$
$$H_{44} = \alpha - \beta$$

The two parts of the original determinant now read:

$$\begin{vmatrix} \alpha - E & \beta \\ \beta & \alpha + \beta - E \end{vmatrix} = 0 \text{ for the energies of the orbitals } \phi_1, \phi_2 \text{ of class } A$$

$$\begin{vmatrix} \alpha - E & \beta \\ \beta & \alpha - \beta - E \end{vmatrix} = 0 \text{ for the energies of the orbitals } \phi_3, \phi_4 \text{ of class } B$$

If we now adopt the usual procedure of dividing through by β and writing

$$\frac{\alpha - E}{\beta} = x$$

the determinants become

$$\begin{vmatrix} x & 1 \\ 1 & x + 1 \end{vmatrix} = 0 \quad \text{or} \quad x^2 + x - 1 = 0$$

$$\begin{vmatrix} x & 1 \\ 1 & x - 1 \end{vmatrix} = 0 \quad \text{or} \quad x^2 - x - 1 = 0$$

the first determinant gives the energies of the A orbitals as

$$x = \frac{-1 \pm \sqrt{5}}{2} = -1 \cdot 618 \text{ or } +0 \cdot 618$$

while the second gives the energies of the B orbitals as

$$x = \frac{1 \pm \sqrt{5}}{2} = -0 \cdot 618 \text{ or } +1 \cdot 618.$$

Thus the two levels of class A have energies $(\alpha + 1 \cdot 618\beta)$ and $(\alpha - 0 \cdot 618\beta)$, while those of class B have energies $(\alpha + 0 \cdot 618\beta)$ and $(\alpha - 1 \cdot 618\beta)$. Note that we do not, as in the ordinary calculation, attempt to associate these energies E_1, E_2, E_3, E_4 with the symmetry-adapted orbitals ϕ_1, ϕ_2, ϕ_3, ϕ_4. This is because these orbitals are not, in general, the true molecular orbitals. They are only the true orbitals if the H_{ik} terms $(i \neq k)$ are zero. The true orbitals, which we shall designate ψ_1, ψ_2, ψ_3, ψ_4, are linear combinations of the symmetry-adapted orbitals. If we wish to determine charge density, bond order or free valence index, we need to obtain the true molecular orbitals. The procedure is now illustrated for butadiene.

Consider first the orbitals of class A. The level of lowest energy is $E_1 = \alpha + 1 \cdot 618\beta$; call the orbital associated with this energy level ψ_1 and let this be some linear combination

$$\psi_1 = \frac{1}{(1 + x^2)^{\frac{1}{2}}} [\phi_1 + x\phi_2] \text{ where } \frac{1}{(1 + x^2)^{\frac{1}{2}}} \text{ is the normalising factor. We now}$$

may solve for x as follows

$$E_1 = \alpha + 1 \cdot 618\beta$$

But
$$E_1 = \int \psi_1 \mathcal{H} \psi_1 \, d\tau$$

$$= \frac{1}{(1 + x^2)} \int (\phi_1 + x\phi_2) \mathcal{H} (\phi_1 + x\phi_2) d\tau$$

$$= \frac{1}{(1 + x^2)} \int [\phi_1 \mathcal{H}\phi_1 + 2x\phi_1 \mathcal{H}\phi_2 + x^2 \phi_2 \mathcal{H}\phi_2] d\tau$$

But
$$\int \phi_1 \mathcal{H}\phi_1 \, d\tau = H_{11} = \alpha$$

$$\int \phi_1 \mathcal{H}\phi_2 \, d\tau = H_{12} = \beta$$

$$\int \phi_2 \mathcal{H}\phi_2 \, d\tau = H_{22} = \alpha + \beta$$

$\therefore$
$$E_1 = \left(\frac{1}{1 + x^2} \right) [\alpha + 2\beta x + x^2 (\alpha + \beta)] = \alpha + 1 \cdot 618\beta$$

$\therefore$
$$\alpha + 2\beta x + x^2 (\alpha + \beta) = (1 + x^2)(\alpha + 1 \cdot 618\beta)$$

The terms in α cancel, leaving

$$2\beta x + x^2\beta = 1.618\beta + 1.618x^2\beta$$

on dividing through by β and rearranging

$$0.618x^2 - 2x + 1.618 = 0$$

from which $x = 1.618$

Thus $\qquad \psi_1 = \dfrac{1}{(1 + 1.618^2)^{\frac{1}{2}}}[\phi_1 + 1.618\,\phi_2]$

$$= \frac{1}{(1 + 1.618^2)^{\frac{1}{2}}}\left[\frac{1}{\sqrt{2}}(\Phi_1 + \Phi_4) + 1.618\cdot 7\,\frac{1}{\sqrt{2}}(\Phi_2 + \Phi_3)\right]$$

$$= 0.376\,(\Phi_1 + \Phi_4) + 0.607\,(\Phi_2 + \Phi_3)$$

The other energy level corresponding to an orbital of class A is $E_3 = \alpha - 0.618\beta$. We may form ψ_3 as the linear combination

$$\psi_3 = \frac{1}{(1 + x^2)^{\frac{1}{2}}}(\phi_1 - x\phi_2)$$

$$E_3 = \int \psi_3 \mathcal{H} \psi_3\, d\tau = \frac{1}{(1 + x^2)}\int(\phi_1 \mathcal{H}\phi_1 - 2x\phi_1 \mathcal{H}\phi_2 + x^2\phi_2 \mathcal{H}\phi_2)d\tau$$

$$= \frac{1}{1 + x^2}[\alpha - 2\beta x + x^2\beta] = \alpha - 0.618\beta$$

Then, proceeding as for E_1 we have

$$-2x + x^2 = -0.618(1 + x^2)$$

$$1.618x^2 - 2x + 0.618 = 0$$

from which $x = 0.618$

and $\qquad \psi_3 = 0.607\,(\Phi_1 + \Phi_4) - 0.376\,(\Phi_2 + \Phi_3)$

For the energy levels associated with orbitals of class B, we have

$$E_2 = \alpha + 0.618\beta;\; \psi_2 = \phi_3 + x\phi_4 \text{ where } x = 0.618$$

giving $\qquad \psi_2 = 0.607\,(\Phi_1 - \Phi_4) + 0.376\,(\Phi_2 - \Phi_3)$

and $\qquad E_4 = \alpha - 1.618\beta;\; \psi_4 = \phi_3 - x\phi_4 \text{ where } x = 1.618$

giving $\qquad \psi_4 = 0.376\,(\Phi_1 - \Phi_4) - 0.607\,(\Phi_2 - \Phi_3)$

For planar cyclic conjugated systems, of which benzene is the best-known, the procedure for determining the energies is slightly altered. The alteration involves omission of the second step; this is because the reducible representation formed by the p-orbitals of such systems contains every irreducible representation of the rotation sub-group C_n once. With the atoms labelled as:

we now carry out the remainder of the procedure for benzene.

1. The point group of benzene is D_{6h} and the p-orbitals are labelled $\Phi_1 - \Phi_6$. The required sub-group of D_{6h} is C_6. Table 7.11 is the character table of the point group C_6.

Table 7.11. Character table of the point group C_6

	I	C_6^1	$C_6^2 \equiv C_3^1$	$C_6^3 \equiv C_2$	$C_6^4 \equiv C_3^2$	C_6^5		
A	1	1	1	1	1	1	R_z	$z, x^2 + y^2, z^2$
B	1	-1	1	-1	1	-1		
E_1 (a)	1	ϵ	$-\epsilon^*$	-1	$-\epsilon$	ϵ^*	(R_x, R_y)	$(x,y)(xz,yz)$
(b)	1	ϵ^*	$-\epsilon$	-1	$-\epsilon^*$	ϵ		
E_2 (a)	1	$-\epsilon^*$	$-\epsilon$	1	$-\epsilon^*$	$-\epsilon$		$(x^2 - y^2, xy)$
(b)	1	$-\epsilon$	$-\epsilon^*$	1	$-\epsilon$	$-\epsilon^*$		

$\epsilon = exp\,(2\pi i/n) = cos2\pi/n + i\,sin2\pi/n$
ϵ^* is the complex conjugate of $\epsilon = cos2\pi/n - i\,sin2\pi/n$ where n is the order of the rotation axis = 6

2. Step 2 is omitted since we know that each irreducible representation occurs once, that is, we require one orbital of class A, one of class B, and one derived from each of the degenerate pairs of E_1 and E_2.

3. To construct the symmetry-adapted orbitals, we need to know what happens to each orbital under the symmetry operations of the group C_6. This information is usually presented as a so-called transformation table and is given here as Table 7.12.

Table 7.12. Transformation table for the atomic p-orbitals of the carbon atoms in benzene

	I	C_6^1	C_3^1	C_2	C_3^2	C_6^5
Φ_1	1	2	3	4	5	6
Φ_2	2	3	4	5	6	1
Φ_3	3	4	5	6	1	2
Φ_4	4	5	6	1	2	3
Φ_5	5	6	1	2	3	4
Φ_6	6	1	2	3	4	5

Using equation 7.1, with Φ_1 as the generating function Φ_G in every case, we obtain the following un-normalised symmetry-adapted orbitals:

$$\phi_1(A) = \Phi_1 + \Phi_2 + \Phi_3 + \Phi_4 + \Phi_5 + \Phi_6$$

$$\phi_2(B) = \Phi_1 - \Phi_2 + \Phi_3 - \Phi_4 + \Phi_5 - \Phi_6$$

$$\phi_3'(E_1)a = \Phi_1 + \epsilon\Phi_2 - \epsilon^*\Phi_3 - \Phi_4 - \epsilon\Phi_5 + \epsilon^*\Phi_6$$

$$\phi_4'(E_1)b = \Phi_1 + \epsilon^*\Phi_2 - \epsilon\Phi_3 - \Phi_4 - \epsilon^*\Phi_5 + \epsilon\Phi_6$$

$$\phi_5'(E_2)a = \Phi_1 - \epsilon^*\Phi_2 - \epsilon\Phi_3 + \Phi_4 - \epsilon^*\Phi_5 - \epsilon\Phi_6$$

$$\phi_6'(E_2)b = \Phi_1 - \epsilon\Phi_2 - \epsilon^*\Phi_3 + \Phi_4 - \epsilon\Phi_5 - \epsilon^*\Phi_6$$

The orbitals ϕ_3', ϕ_4', ϕ_5', ϕ_6' cannot be conveniently used in the calculation of the energy levels because of the complex terms which they contain. However, since ϕ_3', ϕ_4' and ϕ_5', ϕ_6' each constitute a degenerate pair, any linear combination leads to the same energy. The complex terms can be removed if we form four new symmetry-adapted orbitals as follows:

$$\phi_3 = \phi_3' + \phi_4' \qquad \phi_4 = \frac{1}{i}(\phi_3' - \phi_4')$$

$$\phi_5 = \phi_5' + \phi_6' \qquad \phi_6 = \frac{1}{i}(\phi_5' - \phi_6')$$

Since

$$\epsilon = \cos\frac{2\pi}{6} + i\sin\frac{2\pi}{6}, \quad \epsilon^* = \cos\frac{2\pi}{6} - i\sin\frac{2\pi}{6}$$

we have

$$\epsilon + \epsilon^* = 2\cos\frac{2\pi}{6} = 1$$

$$\epsilon - \epsilon^* = 2i\sin\frac{2\pi}{6} = i.\sqrt{3}$$

Thus

$$\phi_3 = 2\Phi_1 + \Phi_2 - \Phi_3 - 2\Phi_4 - \Phi_5 + \Phi_6$$

$$\phi_4 = \sqrt{3}(\Phi_2 + \Phi_3 - \Phi_5 - \Phi_6)$$

$$\phi_5 = 2\Phi_1 - \Phi_2 - \Phi_3 + 2\Phi_4 - \Phi_5 - \Phi_6$$

$$\phi_6 = \sqrt{3}(\Phi_2 - \Phi_3 + \Phi_5 - \Phi_6)$$

On normalising the six orbitals are now

$$\phi_1(A) = \frac{1}{\sqrt{6}}(\Phi_1 + \Phi_2 + \Phi_3 + \Phi_4 + \Phi_5 + \Phi_6)$$

$$\phi_2(B) = \frac{1}{\sqrt{6}}(\Phi_1 - \Phi_2 + \Phi_3 - \Phi_4 + \Phi_5 - \Phi_6)$$

$$\phi_3(E_1) = \frac{1}{\sqrt{12}}(2\Phi_1 + \Phi_2 - \Phi_3 - 2\Phi_4 - \Phi_5 + \Phi_6)$$

$$\phi_{4(E_1)} = \tfrac{1}{2}(\Phi_2 + \Phi_3 - \Phi_5 - \Phi_6)$$

$$\phi_{5(E_2)} = \frac{1}{\sqrt{12}}(2\Phi_1 - \Phi_2 - \Phi_3 + 2\Phi_4 - \Phi_5 - \Phi_6)$$

$$\phi_{6(E_2)} = \tfrac{1}{2}(\Phi_2 - \Phi_3 + \Phi_5 - \Phi_6)$$

We have now broken the energy-level problem down into four smaller problems; there is one orbital of each of the classes A and B, giving linear equations for the corresponding energy levels, and two of classes E_1 and E_2, giving quadratic, equations for the corresponding energy levels.

The energy $E_{(A)}$ associated with the A orbital is given by

$$E_{(A)} = \int \phi_1 \mathcal{H} \phi_1 \, d\tau$$

$$= \frac{1}{\sqrt{6}} \times \frac{1}{\sqrt{6}} \int [\Phi_1 + \Phi_2 + \Phi_3 + \Phi_4 + \Phi_5 + \Phi_6] \mathcal{H} [\Phi_1 + \Phi_2 + \Phi_3 + \Phi_4 + \Phi_5 + \Phi_6] \, d\tau$$

$$= \frac{1}{6}(6\alpha + 12\beta) = \alpha + 2\beta$$

Similarly

$$E_{(B)} = \int \phi_2 \mathcal{H} \phi_2 \, d\tau$$

$$= \frac{1}{\sqrt{6}} \times \frac{1}{\sqrt{6}} \int [\Phi_1 - \Phi_2 + \Phi_3 - \Phi_4 + \Phi_5 - \Phi_6] \mathcal{H} [\Phi_1 - \Phi_2 + \Phi_3 - \Phi_4 + \Phi_5 - \Phi_6] \, d\tau$$

$$= \frac{1}{6}[6\alpha - 12\beta] = \alpha - 2\beta$$

The symmetry-adapted orbitals of classes A and B must be the true molecular orbitals, since the secular equations are simply

$$[H_{11} - E_A] = 0, \quad [H_{22} - E_B] = 0 \text{ respectively}$$

For the orbitals of Class E_1, which are a degenerate pair, we have

$$E_{(E_1)} = \int \phi_3 \mathcal{H} \phi_3 \, d\tau = \int \phi_4 \mathcal{H} \phi_4 \, d\tau$$

$$\int \phi_3 \mathcal{H} \phi_3 \, d\tau = \frac{1}{\sqrt{12}} \times \frac{1}{\sqrt{12}} \int [2\Phi_1 + \Phi_2 - \Phi_3 - 2\Phi_4 - \Phi_5 + \Phi_6]$$

$$\mathcal{H} [2\Phi_1 + \Phi_2 - \Phi_3 - 2\Phi_4 - \Phi_5 + \Phi_6] \, d\tau$$

$$= \frac{1}{12}[12\alpha + 12\beta] = \alpha + \beta$$

The reader may verify that $\int \phi_4 \mathcal{H} \phi_4 \, d\tau = \tfrac{1}{4}(4\alpha + 4\beta) = \alpha + \beta$ also.

Further, we may calculate the value of the off-diagonal element $H_{34} = \int \phi_3 \mathcal{H} \phi_4 \, d\tau =$

$$\frac{1}{\sqrt{12}} \times \frac{1}{2} \int [2\Phi_1 + \Phi_2 - \Phi_3 - 2\Phi_4 - \Phi_5 + \Phi_6] \mathcal{H} [\Phi_2 + \Phi_3 - \Phi_5 - \Phi_6] \, d\tau$$

SYMMETRY AND STEREOCHEMISTRY

and we may conveniently tabulate the products as shown:

ϕ_4 \ ϕ_3	$2\Phi_1$	Φ_2	$-\Phi_3$	$-2\Phi_4$	$-\Phi_5$	Φ_6
Φ_2	2β	α	β	0	0	0
Φ_3	0	β	$-\alpha$	-2β	0	0
$-\Phi_5$	0	0	0	2β	α	$-\beta$
$-\Phi_6$	-2β	0	0	0	β	$-\alpha$

From the previous table we see that the value of H_{34} is zero. This means that the two symmetry-adapted orbitals ϕ_3, ϕ_4 of class E_1 are also true molecular orbitals ψ_3, ψ_4. Finally, for the energies of the E_2 orbitals we have

$$E_{(E_2)} = \int \phi_5 \mathcal{H} \phi_5 \, d\tau = \int \phi_6 \mathcal{H} \phi_6 \, d\tau$$

Taking $\int \phi_6 \mathcal{H} \phi_6 \, d\tau$ this time, we have

$$E_{(E_2)} = \tfrac{1}{2} \times \tfrac{1}{2} \int [\Phi_2 - \Phi_3 + \Phi_5 - \Phi_6] \, \mathcal{H} [\Phi_2 - \Phi_3 + \Phi_5 - \Phi_6] \, d\tau$$
$$= \tfrac{1}{4}[4\alpha - 4\beta] = \alpha - \beta$$

The reader may verify, firstly, that $\int \phi_5 \mathcal{H} \phi_5 \, d\tau = \frac{1}{12}(12\alpha - 12\beta) = \alpha - \beta$ also, and

secondly that $\int \phi_5 \mathcal{H} \phi_6 \, d\tau = 0$, showing that the symmetry-adapted orbitals ϕ_5, ϕ_6

are the true orbitals ψ_5, ψ_6.

It is perhaps of interest to determine the irreducible representations of the full point group, D_{6h}, to which these orbitals belong. While it is possible to do this simply by constructing the character of the reducible representation of the p-orbitals for the full point group, and breaking this down into its component irreducible representations, it is not necessary to do this. First of all we know that all the orbitals must be antisymmetric with respect to the operation σ_h since this reverses the signs of the lobes above and below the molecular plane. Thus the irreducible representation to which each orbital belongs must be one in which the character of σ_h is negative. For the E orbitals, this immediately shows that the E_1 orbitals belong to E_{1g} and the E_2 to E_{2u}. Also the A orbital must be either A_{1u} or A_{2u} and the B orbital either B_{1g} or B_{2g}. We can make this distinction by considering the behaviour of the orbitals under reflection in one of the vertical planes which pass through the opposite atoms. Consider the reflection in the plane passing through atoms 1 and 4. This will interchange atoms 2 and 6, 3 and 5. Since the p-orbital itself is symmetric to such a reflection, we need to consider only the signs of the coefficients in the orbital. Interchange of Φ_2 with Φ_6 and Φ_3 with Φ_5 leaves ϕ_1 unaltered; its character to σ_v is therefore +1 and it therefore

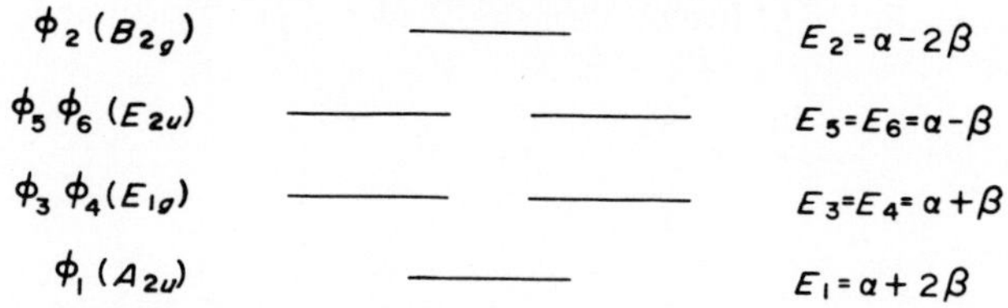

$$\phi_2\ (B_{2g}) \qquad\qquad ——— \qquad\qquad E_2 = \alpha - 2\beta$$

$$\phi_5\ \phi_6\ (E_{2u}) \qquad ——— \quad ——— \qquad E_5 = E_6 = \alpha - \beta$$

$$\phi_3\ \phi_4 (E_{1g}) \qquad ——— \quad ——— \qquad E_3 = E_4 = \alpha + \beta$$

$$\phi_1\ (A_{2u}) \qquad\qquad ——— \qquad\qquad E_1 = \alpha + 2\beta$$

Figure 7.11   Energy levels in benzene.

belongs to the irreducible representation A_{2u}. In ϕ_2, also, these interchanges leave the orbital unaltered, showing that ϕ_2 belongs to B_{2g}. Figure 7.11 shows the energies and symmetries of the molecular orbitals of benzene.

Conservation of Orbital Symmetry in Chemical Reaction

Up to now we have been considering descriptions of orbitals in terms of symmetry which enable us to explain the structures and physical properties of molecules. It is, however, possible to make use of symmetry arguments to explain and predict the course of certain chemical reactions. The *Woodward-Hoffman* approach shows how the need for conservation of orbital symmetry between reactants and products in processes such as cyclo-addition and electro-cyclic reactions can be used to formulate acceptable hypotheses concerning the nature of the reaction paths involved. Let us illustrate this approach with respect to the following four reactions:

(a) The ring opening reaction of cyclobutene

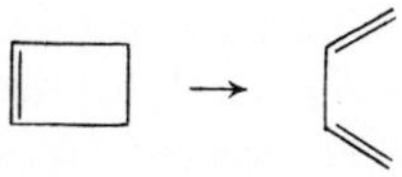

(b) The electrocyclic reaction of hexa-1,3,5-triene

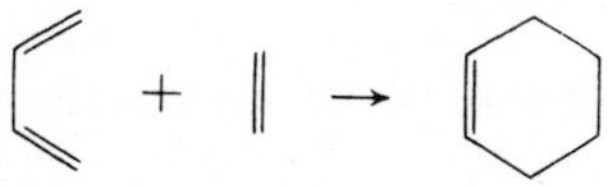

(c) The cyclo-addition of *cisoid*-buta-1,3,-diene to ethylene

(d) The cyclo-addition of two ethylene molecules to give cyclobutane

We can consider the approach as involving four stages:

1. Determination of the symmetry of the molecular orbitals of the reactants likely to be involved in the reaction with respect to the symmetry elements present in the molecules. We do not have to consider the symmetry of the orbitals with respect to all of the elements present, but only with respect to those which bisect the bonds being broken in a cyclo-reversion or formed during a cyclo-addition.

2. Determination of the symmetry of the relevant molecular orbitals of the products, i.e. these orbitals which arise as a result of the reaction. The symmetry considerations are the same as those of the previous stage.

3. Construction of a *Correlation Diagram* for the molecular orbitals of the reactants and products.

4. Determination from the correlation diagram whether there is orbital conservation in the ground states or the excited states of the reactants or products. From this information we can predict acceptable reaction paths.

In the reaction (a) the reactant is cyclobutene. In order to form *cisoid*-buta-1,3-diene from this we must break one π and one σ bond and form two new π bonds. The orbitals involved in the bond breaking and formation are shown in Figure 7.12, which also shows the order of energy levels for the possible combinations of these orbitals. In the ground states of the molecules the electronic configurations would be those indicated. Figure 7.13 shows the symmetry elements of the reactant and product molecules with respect to which the orbital symmetry is to be considered, i.e. the elements (C_2, σ_v) which bisect the bonds being broken in this cyclo-reversion. Table 7.13 shows the symmetry of the orbitals of the reactant and product with respect to C_2 and σ_v.

Table 7.13. Symmetry of cyclobutene and buta-1,3-diene orbitals with respect to C_2 and σ_v

cyclobutene				cisoid-buta-1,3-diene		
	C_2	σ_v			C_2	σ_v
σ^*	A	A		ϕ_4	S	A
π^*	S	A		ϕ_3	A	S
π	A	S		ϕ_2	S	A
σ	S	S		ϕ_1	A	S

S = *molecular orbital symmetric with respect to the symmetry element*
A = *molecular orbital antisymmetric with respect to the symmetry element.*

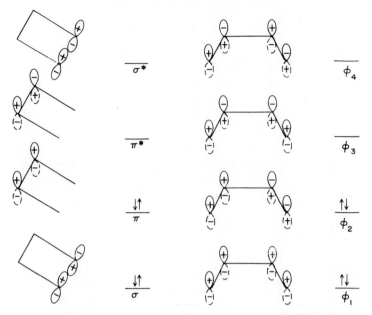

Figure 7.12   Orbitals involved in the reaction cyclobutene ⇌ *cisoid*-buta-1,3-diene and
                   the energy level diagrams for the molecular orbitals in the isolated molecules

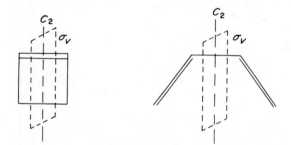

Figure 7.13   The C_2 and σ_v symmetry elements which bisect the bond broken in the
                   cycloreversion reaction cyclobutene ⇌ *cisoid*-buta-1,3-diene

This information enables us to construct the correlation diagrams of Figure
7.14. A correlation diagram is constructed by first listing the relevant molecular
orbitals of the reactants and products, taking into account the approximate
energies of the various orbitals, and then joining reactant and product levels
which have the same symmetry. The approximate energies of the levels of
product and reactant cannot be obtained from symmetry considerations but
must be estimated by other means (e.g. Hückel and related calculations). We
can see from Table 7.13 that we cannot, if orbital symmetry is to be conserved

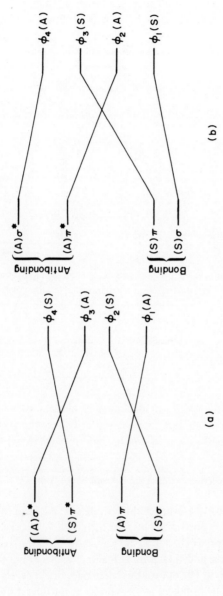

Figure 7.14   Correlation diagrams for the formation of *cisoid*-buta-1,3-diene from cyclobutene preserving (a) (conrotatory) the C_2 axis (b) (disrotatory) the mirror plane through the cyclobutene bond being broken

construct a correlation diagram to represent the cyclo-reversion of cyclobutene via a transition state which preserves both of the symmetry elements C_2 and σ_v. We cannot construct such a diagram because we do not have product orbitals with SS and AA symmetry with respect to the elements to match those of the reactant. That these observations must be correct can be seen from Figure 7.15 which shows how the σ-bond in cyclobutene can be broken. To break this bond and put orbitals in the correct positions for bond formation in *cisoid*-buta-1,3,-diene they must be rotated out of the plane of the cyclobutene ring containing the bond in one of the ways shown. The conrotatory mode preserves the C_2 axis through the cyclobutene bond being broken while the disrotatory mode retains the mirror plane (see Figure 7.15). In the conrotatory mode electrons from the

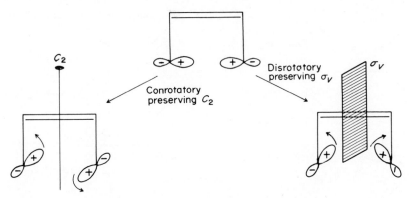

Figure 7.15   Conrotatory and disrotatory modes of bond breaking in cyclobutene

ground state in cyclobutene would form the ground state electrons in *cisoid*-buta-1,3-diene and such a reaction can result from a thermal process. For a reaction path to involve the disrotatory mode, however, there must be some population of the $A(\pi^*)$ level of the reactant in order to obtain the ground state configuration of the product. Such a reaction can only occur with the excited state of cyclobutene and thus results from a photochemical process.

In the reaction (b) the reactant is hexa-1,3,5-triene and the product cyclo-hexa-1,3-diene. In this ring closure reaction, we must consider the symmetries of the molecular orbitals in Figure 7.16 with respect to the symmetry elements (C_2 and σ_v) which bisect the bond being formed in cyclohexa-1,3-diene. Again we cannot construct a correlation diagram to represent a transition state in which both C_2 and σ_v are preserved. We can, however, construct the correlation diagrams of Figure 7.17 in which (a) C_2 is preserved (conrotatory) and (b) σ_v is preserved (disrotatory). In this reaction the mode which involves electrons in ground state orbitals of the reactants is the disrotatory mode which must therefore represent the thermal process. The conrotatory mode requires the

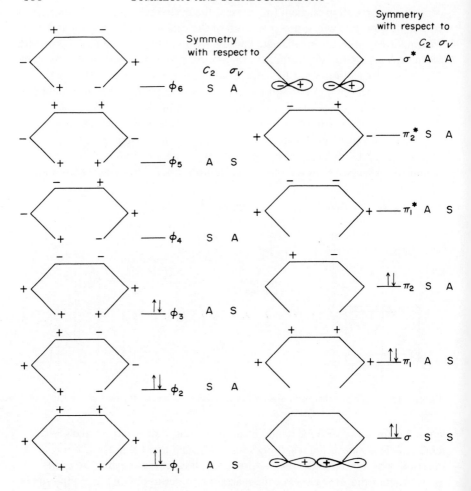

Only the sign of the upper lobe of each
π - bond forming orbital is shown

Figure 7.16   Orbitals involved in the reaction hexa-1,3,5-triene $\rightleftharpoons$ cyclohexa-1,3-diene,
showing their occupation in the isolated molecules and their symmetry
with respect to the elements C_2 and σ_v bisecting the bond to be formed

population of the antibonding orbital (ϕ_4) and must represent the photochemical
process.

We may observe that in reaction (a) the thermal reaction involves the conrotatory mode while in reaction (b) it involves the disrotatory mode. The obvious
difference in the two reactions lies in the number of electrons involved in the
process, 4 electrons in (a), 6 in (b). The Woodward-Hoffman approach shows how

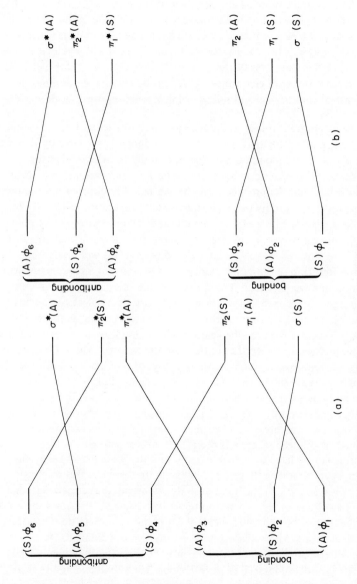

Figure 7.17   Correlation diagrams for the formation of cyclohexa-1,3-diene from hexa-1,3,5-triene preserving (a) the C_2 axis and (b) the mirror plane through the bond being formed

the number of electrons involved in a reaction determines the rotary modes of the thermal and photochemical processes. The correlation diagrams for the conrotatory mode are always of the same type in that they involve crossing over in pairs starting from the lowest level. This is illustrated in Figure 7.18 a which is a generalised diagram showing crossing over of the type $1\rightarrow2$, $2\rightarrow1$, $3\rightarrow4$, $4\rightarrow3$, etc. Likewise the diagrams for the disrotatory mode are alike in that the first crossover is always $1\rightarrow1$ and then they involve crossing over in pairs starting from the second level. This is illustrated in Figure 7.18(b) which is a generalised diagram showing crossing over of the type $1\rightarrow1$, $2\rightarrow3$, $3\rightarrow2$, $4\rightarrow5$, $5\rightarrow4$, etc. We can see how the number of electrons involved in the reaction determines the rotary sense of the thermal and photochemical reactions by considering a few examples.

If two electrons are involved in the reaction, only the levels 1 and 2 of the generalised diagrams of Figure 7.18 need be considered. Level 1 would then be bonding and level 2, antibonding. For disrotation the transfer of electrons between reactant and product orbitals would be $1\rightarrow1$, only ground state orbitals would be involved and the process would be thermal. For conrotation the transfer of electrons to the product ground state would proceed via level 2 of the reactant (i.e. an excited state of the reactant) and the process would be photochemical. If four electrons are involved in the reaction, levels 1, 2, 3 and 4 of Figure 7.18 are required. Levels 1 and 2 would be bonding and levels 3 and 4 antibonding. Reaction (a) described earlier is an example of a process involving four electrons and we have seen that in this case the thermal reaction is conrotatory and the photochemical reaction disrotatory. If six electrons are involved in the reaction, levels 1 to 6 are required, 1 to 3 being bonding and 4 to 6 being antibonding. Reaction (b) described earlier is an example of a process involving six electrons and we have seen that in this case the thermal reaction is disrotatory and the photochemical reaction conrotatory. Figure 7.18 can be used to work out the rotatory sense of the thermal and photochemical processes for reactions involving any even number of electrons and the result can be expressed in the following generalisation. If there are $2n$ electrons involved where n is odd the thermal process is disrotatory and the photochemical process conrotatory but if $2n$ electrons are involved where n is even the converse is true.

In reaction (c) we have two reactants, *cisoid*-buta-1,3-diene and ethylene, involved in a cyclo-addition to give cyclohexene. In cases like this we must consider the ways in which the bond orbitals can orientate themselves in the reactant molecules in order to permit interactions. Figure 7.19 shows the ways in which the outermost orbitals forming the π-bonds can orientate themselves to enable a reaction with similar orbitals on ethylene to take place. The two possible modes of orientation of orbitals on a single reactant are suprafacial (Figure 7.19 a) and antarafacial (Figure 7.19 b). The suprafacial mode clearly preserves a mirror plane while the antarafacial mode preserves a C_2 axis. If we consider two reactants there are four possible combinations of suprafacial (s) and

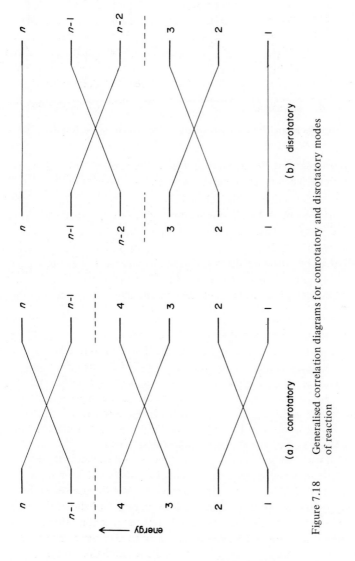

(a) conrotatory   (b) disrotatory

Figure 7.18   Generalised correlation diagrams for conrotatory and disrotatory modes of reaction

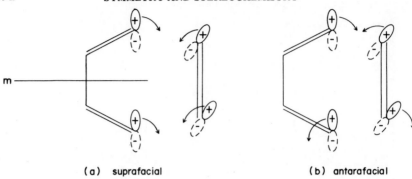

(a) suprafacial   (b) antarafacial

Figure 7.19   Suprafacial and antarafacial orientation of orbitals on *cisoid*-buta-1,3-diene

antarafacial (a) orientations viz. ss, as, sa and aa. If the two reactions happen to be identical (e.g. ethylene and ethylene) then we cannot distinguish sa and as.

For reaction (c) we can construct the correlation diagrams (Figure 7.20) for the cases where a mirror plane and a C_2 axis are conserved throughout the reaction, i.e. for the ss and aa cases. These correlation diagrams have the molecular orbitals for both reactants on the left and those for the product on the right. In reactions involving more than one reactant the correlation diagram has to be constructed with respect to a symmetry element linking all of the reactants and the product. In the reaction between *cisoid*-buta-1,3-diene and ethylene we can see that the

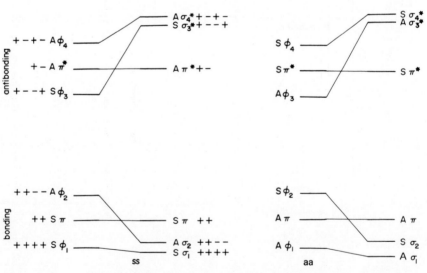

Figure 7.20   Correlation diagrams for the reaction *cisoid*-buta-1,3-diene + ethylene $\rightleftharpoons$ cyclohexene

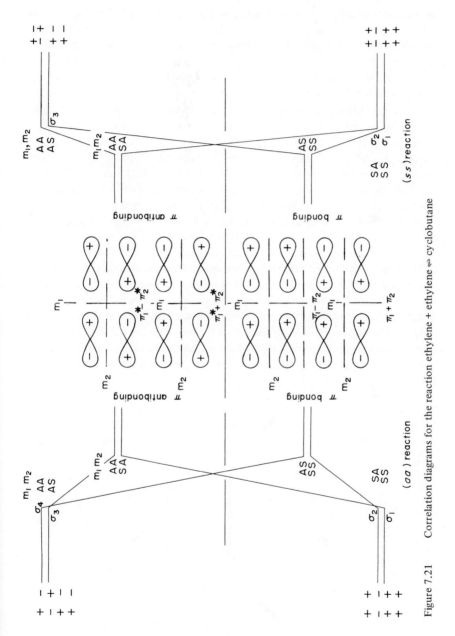

Figure 7.21   Correlation diagrams for the reaction ethylene + ethylene ⇌ cyclobutane

correlation diagrams for the ss and aa cases are similar, in that bonding levels of the reactants correspond to the bonding levels of the product. There is no correlation which crosses the bonding-antibonding energy barrier (see Figure 7.20). Since the reactant orbitals required are ground state orbitals the reaction is symmetry allowed in the ground state. If an electron were to be promoted to the first excited state the reaction would become symmetry forbidden because of the absence of a bonding-antibonding crossover.

In reaction (d) we have two reactants, both of which are ethylene molecules, involved in a cyclo-addition to give cyclobutane. Again we can construct the correlation diagrams for the ss and aa cases. These diagrams, which are shown in Figure 7.21, differ from those of Figure 7.20 in that there is a crossover between reactant antibonding levels and product bonding levels. This means that, if the reactant electrons remain in their ground state orbitals, reaction cannot occur and thus the reaction is symmetry forbidden in the ground state. The reaction is, however, symmetry allowed in the excited state and will proceed if a reactant electron is promoted to an antibonding molecular orbital.

So far we have only considered reactions involving the most obvious approach of two reactants. This has allowed us to construct correlation diagrams for the aa and ss cases. For some reactions it is possible to construct correlation diagrams for sa and as modes of interaction provided we can work out a suitable geometry of approach. For example this can be done for the ethylene + ethylene reaction provided the two molecules approach orthogonally (Figure 7.22). There is, however, no suitable mode of approach which would enable us to construct meaningful correlation diagrams for the sa and as modes in the ethylene-*cisoid*-buta-1,3-diene reaction.

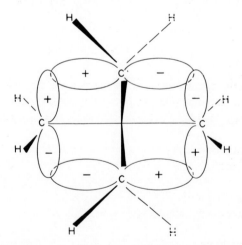

Figure 7.22   Orthogonal approach of two ethylene molecules permitting an *as* mode of reaction

The Woodward-Hoffman approach again shows how the numbers of electrons involved in a reaction can be used to determine whether a cycloaddition process is symmetry allowed in the ground state or the excited state. Processes involving ss or aa modes are symmetry allowed in the ground state if they involve a total of $2n$ electrons when n is odd and symmetry allowed in the excited state if they involve a total of $2n$ electrons when n is even.

PROBLEMS

1. Show by considering the effect of the operation of the symmetry elements of the appropriate point group that

(a) p_x has symmetry B_1 in C_{2v} (character table page 154)

(b) d_{z^2} has symmetry A_1 in C_{3v} (character table page 154)

(c) $d_{x^2-y^2}$ has symmetry B_{1g} in D_{4h} (character table page 168)

(d) p_z has symmetry A_2 in D_3 (character table page 221)

(e) d_{xy} has symmetry B_2 in C_{4v} (character table page 222)

(f) f_{xyz} has symmetry A_{2u} in O_h (character table page 155)

The f_{xyz} orbital has the following shape

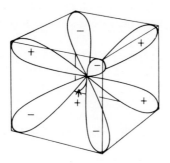

2. By reference to the appropriate character tables (see appendix II) explain the likely splittings of p- and d-orbitals in electrostatic fields of symmetry

(a) D_{4h} (b) D_{4d} (c) D_3 (d) D_{3h} (e) D_{2d} (f) C_4 (g) C_{2v}

3. Work out the total character of the σ bond orbitals in the following molecules and list the atomic orbitals of the central atom which would contribute to the hybrid.

(a) SF_6 with O_h symmetry.

(b) XeF_4 with D_{4h} symmetry.

4. The seven f orbitals are f_{xyz}, $f_{x(y^2-z^2)}$, $f_{y(z^2-x^2)}$, $f_{z(x^2-y^2)}$, f_{x^3}, f_{y^3} and f_{z^3}. If the following functions transform in O_h symmetry xyz, as A_{2u};

$(x^3, y^3, z^3), T_{1u}, \{x(y^2 - z^2), y(z^2 - x^2), z(x^2 - y^2)\}, T_{2u}$; how would the f orbitals be split in an octahedral field?

5. Show that the energy levels of the allyl system are $\alpha + \sqrt{2}\beta, \alpha, \alpha - \sqrt{2}\beta$ and obtain the true molecular orbitals.

6. Find the energy-levels and true molecular orbitals for cyclopropene assuming (i) an equilateral triangular shape and (ii) an isosceles triangular shape for the molecule.

7. Verify, by determining the character of the reducible representation formed by the p-orbitals forming the conjugated system in benzene, that, under the full point group symmetry (D_{6h}) the orbitals are $A_{2u} + B_{2g} + E_{1g} + E_{2u}$.

Appendix I Answers to Problems

Chapter 1

1. (a) rectangular (b) diamond shaped

2. (i) (1 2 2) (ii) (4 3 6)
   (iii) (2 1 $\bar{1}$) (iv) (4 $\bar{2}$ 1)

3. The diagram should be similar to that of Figure 1.6. The (3 2 2) plane can be illustrated by drawing the triangle formed by joining points which are $\frac{1}{3}a$ along x, $\frac{1}{2}b$ along y and $\frac{1}{2}c$ along z. The (3 2 2) plane is similarly illustrated by joining the points which are $\frac{1}{3}a$ along x, $\frac{1}{2}b$ along y and $\frac{1}{2}c$ along z.

4. (1 1 1), ($\bar{1}$ 1 1), ($\bar{1}$ $\bar{1}$ 1), etc.

Chapter 2

1. The lines joining the centres of opposite faces are 4-fold and 2-fold axes. Lines coincident with the square diagonals are 2-fold axes.

2. (a) The molecular plane is a σ_v.
   (b) The molecular plane and the plane perpendicular to it, bisecting the $\hat{HOH}$ angle are both σ_v.
   (c) The molecular plane is a σ_h, and the three planes perpendicular to it, each containing an N–O bond, are $3\sigma_v$.
   (d) The molecular plane is a σ_h; there are four σ perpendicular to it. Two of these bisect an $\hat{FXeF}$ angle and are designated σ_d; the other two each contain two Xe-F bonds and are σ_v.
   (e) Any plane containing the carbon and two of the chlorine atoms is a σ_d. There are six of these.
   (f) There are three σ_h containing the sulphur and four fluorine atoms, and six σ_d which each bisect a pair of vertically opposite $\hat{FSF}$ angles.

3. (a) The internuclear axis is a C_∞.

(b) The internuclear axis is a C_∞ and any line through the carbon atom, perpendicular to this axis, is a C_2. There is an infinite number of C_2 axes.

(c) The line bisecting the $\widehat{OSO}$ angle is a C_2.

(d) Each line bisecting a $\widehat{CCC}$ angle is a C_2. There are 5 of these. The line through the centre of the anion perpendicular to the plane in which it lies is a C_5.

(e) C_2 axes lie in the Cl–Pt–Cl directions and also in the direction of the lines bisecting $\widehat{ClPtCl}$ angles. The line through the Pt atom perpendicular to the plane in which the anion lies is a C_4 and a C_2.

(f) The C–C direction is a threefold axis. There are three C_2 axes whose directions are shown in Figure 3.15.

4. (a) (b) (c) (d) (f) (g).

5. (a)

	I	C_2	σ_v	σ_v'
I	I	C_2	σ_v	σ_v'
C_2	C_2	I	σ_v'	σ_v
σ_v	σ_v	σ_v'	I	C_2
σ_v'	σ_v'	σ_v	C_2	I

σ_v is the molecular plane

(b)

	I	C_4^1	C_4^3	$C_4^2 \equiv C_2$	σ_{v1}	σ_{v2}	σ_{d1}	σ_{d2}
I	I	C_4^1	C_4^3	C_2	σ_{v1}	σ_{v2}	σ_{d1}	σ_{d2}
C_4^1	C_4^1	C_2	I	C_4^3	σ_{d1}	σ_{d2}	σ_{v2}	σ_{v1}
C_4^3	C_4^3	I	C_2	C_4^1	σ_{d2}	σ_{d1}	σ_{v1}	σ_{v2}
$C_4^2 = C_2$	C_2	C_4^3	C_4^1	I	σ_{v2}	σ_{v1}	σ_{d2}	σ_{d1}
σ_{v1}	σ_{v1}	σ_{v2}	σ_{d1}	σ_{v2}	I	C_2	C_4^3	C_4^1
σ_{v2}	σ_{v2}	σ_{d1}	σ_{d2}	σ_{v1}	C_2	I	C_4^1	C_4^3
σ_{d1}	σ_{d1}	σ_{v1}	σ_{v2}	σ_{d2}	C_4^1	C_4^3	I	C_2
σ_{d2}	σ_{d2}	σ_{v2}	σ_{v1}	σ_{d1}	C_4^3	C_4^1	C_2	I

With the labelling shown in the figure on page 27.

(c)

	I	$C_2(z)$	$C_2(y)$	$C_2(x)$	i	$\sigma_v(xy)$	$\sigma_v(xz)$	$\sigma_v(yz)$
I	I	$C_2(z)$	$C_2(y)$	$C_2(x)$	i	$\sigma_v(xy)$	$\sigma_v(xz)$	$\sigma_v(yz)$
$C_2(z)$	$C_2(z)$	I	$C_2(x)$	$C_2(y)$	$\sigma_v(xy)$	i	$\sigma_v(yz)$	$\sigma_v(xz)$
$C_2(y)$	$C_2(y)$	$C_2(x)$	I	$C_2(z)$	$\sigma_v(xz)$	$\sigma_v(yz)$	i	$\sigma_v(xy)$
$C_2(x)$	$C_2(x)$	$C_2(y)$	$C_2(z)$	I	$\sigma_v(yz)$	$\sigma_v(xz)$	$\sigma_v(xy)$	i
i	i	$\sigma_v(xy)$	$\sigma_v(xz)$	$\sigma_v(yz)$	I	$C_2(z)$	$C_2(y)$	$C_2(x)$
$\sigma_v(xy)$	$\sigma_v(xy)$	i	$\sigma_v(yz)$	$\sigma_v(xz)$	$C_2(z)$	I	$C_2(x)$	$C_2(y)$
$\sigma_v(xz)$	$\sigma_v(xz)$	$\sigma_v(yz)$	i	$\sigma_v(xy)$	$C_2(y)$	$C_2(x)$	I	$C_2(z)$
$\sigma_v(yz)$	$\sigma_v(yz)$	$\sigma_v(xz)$	$\sigma_v(xy)$	i	$C_2(x)$	$C_2(y)$	$C_2(z)$	I

The axis perpendicular to the molecular plane is the z-axis and that through the two chlorine atoms is the y-axis.

Chapter 3

1. (a) The S_6 axis lies in the direction of the C–C bond. It is also a C_3 axis.
(b) The S_2 axis is perpendicular to the molecular plane and passes through the mid-point of the central C–C bond.
(c) The S_3 axis passes through the boron atom and is perpendicular to the molecular plane. It is also a C_3 axis.
(d) The lines joining opposite fluorine atoms are S_4 axes; they are also C_4 and C_2. The lines joining the mid-points of the triangular faces of the SF_6 octahedron are S_6 axes and also C_3 axes.

2. $\bar{8} = S_8 \qquad \bar{10} = S_5$

3. (a) The equivalence of operations between C_{5h} and S_5 is as follows:
$S_5^1 = C_5^1 \times \sigma_h$; $S_5^2 = C_5^2$; $S_5^3 = C_5^3 \times \sigma_h$; $S_5^4 = C_5^4$; $S_5^5 = \sigma_h$;

$S_5^6 = C_5^1$; $S_5^7 = C_5^2 \times \sigma_h$; $S_5^8 = C_5^3$; $S_5^9 = C_5^4 \times \sigma_h$; $S_5^{10} = I$
So every operation of S_5 is equivalent to an operation or combination of operations of C_{5h}.
(b) The equivalence of operations between C_{5i} and S_{10}.

$S_{10}^1 = C_5^4 \times i \qquad\qquad S_{10}^6 = C_5^3$
$S_{10}^2 = C_5^1 \qquad\qquad\quad S_{10}^7 = C_5^2 \times i$
$S_{10}^3 = C_5^3 \times i \qquad\qquad S_{10}^8 = C_5^4$
$S_{10}^4 = C_5^2 \qquad\qquad\quad S_{10}^9 = C_5^1 \times i$
$S_{10}^5 = i \qquad\qquad\qquad S_{10}^{10} = I$

(c) The equivalence of operations between C_{2i} and C_{2h}.
C_{2i} consists of the rotation operations C_2^1, C_2^2 and their products with i;
C_{2h} consists of the rotation operations C_2^1, C_2^2 and their products with σ_h
$C_2^1 \times i = \sigma_h = C_2^2 \times \sigma_h$
$C_2^2 \times i = i = C_2^1 \times \sigma_h$
(d) $S_3^1 = C_3^1 \times \sigma_h$; $S_3^2 = C_3^2$; $S_3^3 = \sigma_h$; $S_3^1 \times \sigma_h = C_3$; $S_3^2 \times \sigma_h = C_3^2 \times \sigma_h$;
$S_3^3 \times \sigma_h = I.$

4. There are two possibilities. For n odd, addition of $n\sigma_v$ planes to the S_n group gives the same set of operations as the addition of these planes to a C_{nh} group since $C_{nh} = S_n$ when n is odd. The intersection of each σ_v with the σ_h of C_{nh} gives a C_2 axis; thus we have a C_n axis with nC_2 axes perpendicular to it. Together with the planes of symmetry these elements make up a D_{nh} group. Thus, for example,

$$S_{3v} = D_{3h}$$

For n even, the elements of an S_{nv} group are one S_n, one $C_{(n/2)}$ coincident with it, $n/2C_2$, $n\sigma_d$, and a centre of symmetry if $n/2$ is odd. This is the same set of elements as that which constitutes the point group $D_{(n/2)d}$. Thus, for n even, $S_{nv} = D_{(n/2)d}$; for example, $S_{6v} = D_{3d}$.

5. Because one of its elements is an S_8 axis.

6. $D_{\infty h}$: O; D_{4h}: X; D_{3h}: Y; D_{2h}: H, I; C_{2h}: N,S,Z; C_{2v}: A, B, C, D, K, L, M, T, U, V, W; C_s; F, G, J, P, Q, R.

7. (a) $D_{\infty h}$; (b) C_s; (c) D_{2h}; (d) C_s; (e) C_s; (f) C_{2v}.

8. Tetrahedron, T_d; trigonal biprism, D_{3h}; square-based pyramid, C_{4v}; trigonal prism, D_{3h}; octahedron, O_h; pentagonal-based pyramid, C_{5v}; pentagonal biprism, D_{5h}; hexagonal-based pyramid, C_{6v}; cube, O_h; hexagonal bipyramid, D_{6h}.

9. (a) $D_{\infty h}$; (b) D_{3h}; (c) C_{3v}; (d) T_d; (e) D_{4h}; (f) D_{3h}; (g) C_{2v}; (h) O_h; (i) C_{4v}.

10. C_s; C_{2v}; C_{3v}; C_{2v}; C_{2v}; C_{2h}; C_{2v}; C_{2v}; C_{2v}; C_{2v}; C_s; C_{2v}; D_{2h}; C_s; D_{3d}; C_{2v}.

11. (a) $C_{\infty v}$; (b) C_{2v}; (c) C_s; (d) C_{3v}; (e) C_{2v}; (f) C_{3v} for Y in an axial position, C_{2v} for Y in an equatorial position; (g) C_s for Y in an axial position or an equatorial position; (h) C_{4v}; (i) C_{4v} if Y is *trans* to the lone pair, C_s if Y is *cis* to the lone pair.

12. (a) $D_{\infty h}$; (b) C_{2v}; (c) C_s; (d) C_{2v}; (e) D_{2h} if Ys are *trans*, C_{2v} if they are *cis*; (f) Both Y axial, D_{3h}; both Y equatorial, C_{2v}; one Y axial and one equatorial, C_s; (g) Both Y axial, C_{2v}; both Y equatorial, C_{2v}; one Y axial and one equatorial, C_1; (h) Y *trans* to each other, D_{4h}; *cis*, C_{2v}; (i) Both Y *cis* to lone-pair, and to each other, C_s, both Y *cis* to lone-pair and *trans* to each other, C_{2v}, one Y *cis* and one Y *trans* to lone-pair, C_s.

13.

C_{5v}

D_7

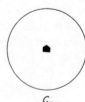

C_{5h}

14. These to be worked out as shown in Figure 3.35.

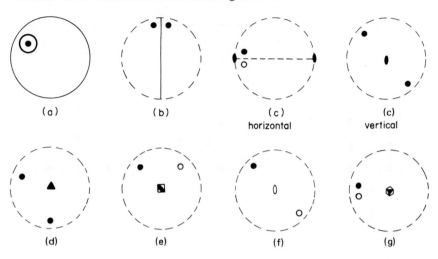

| (a) | (b) | (c)
horizontal | (c)
vertical |

| (d) | (e) | (f) | (g) |

Chapter 4

2. (a) Orthorhombic, primitive cell, axial glide planes (glides $c/2$) perpendicular to both a- and b-axes, c-axis is a 2-fold rotation axis.

(b) Tetragonal, primitive cell, unique axis is a 4-fold rotation axis, axial glide planes (glides $c/2$) perpendicular to the 100 and 110 directions.

(c) Rhombohedral, primitive cell, unique axis is a 3-fold rotation axis.

(d) Orthorhombic, face-centred cell centred on one face, the A-face, mirror plane normal to the a- axis, an axial glide plane (glide $a/2$) perpendicular to the b- axis, the c- axis is a 2-fold axis.

(e) Cubic, face-centred cell, centred on all faces, the 100 directions are 4-fold axes, the 110 directions are 3-fold axes and the 111 directions 2-fold axes.

(f) Monoclinic, primitive cell, the unique axis is a 2_1 screw axis with a mirror plane normal to it.

(g) Hexagonal, primitive cell, the unique axis is a 6_3 screw axis, the other important directions in the cell (see page 93) are all 2-fold axes.

(h) Tetragonal, body-centred cell, the unique axis is a 4_1 screw-axis, there are axial glide-planes (glide $c/2$) perpendicular to the a and b axes and diamond glide planes $\left(\text{glide } \dfrac{a+c}{4}\right)$ perpendicular to the square face diagonals.

3.

pl   pm   pg

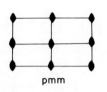
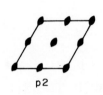

pmm   p2   p4

4.

pl   pm   pg

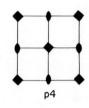

pmm   p2   p4

5.   (i) (a) $x,0,z; \bar{x},0,\bar{z}; x,0,\bar{z}; \bar{x},0,z$
          (b) $x,y,0; \bar{x},\bar{y},0; \bar{x},y,0; x,\bar{y},0$
     (ii) (a) $\frac{1}{2},\frac{1}{2},z; \frac{1}{2},\frac{1}{2},\bar{z}$.   (b) $\frac{1}{2},y,0; \frac{1}{2},\bar{y},0$
    (iii) $0,0,0$.
    (iv) $\frac{1}{4},\frac{1}{4},\frac{1}{4}; \frac{3}{4},\frac{3}{4},\frac{1}{4}; \frac{3}{4},\frac{1}{4},\frac{3}{4}; \frac{1}{4},\frac{3}{4},\frac{3}{4}$
     (v) $0,0,0; \frac{1}{2},\frac{1}{2},\frac{1}{2}$
    (vi) $x,0,z; \bar{x},0,z; 0,x,z; 0,\bar{x},z;$
   (vii) $0,0,z$

6. The symbols $P2_1 22$ represent an orthorhombic space group and can be converted to $P222_1$ by simply changing the order of the axes from *a, b, c* to *b, c, a*. P312 and P321 represent rhombohedral space groups with 3-fold axes in the unique direction. They differ because P312 has 1-fold symmetry about the *a* and *b* axes and 2-fold symmetry about the axes normal to *a, b* and in the 0001 plane while P321 has 2-fold symmetry about the *a*-and *b*-axes and 1-fold symmetry along the other specified axes.

7. (a), (b), (d).

Chapter 5

1. The products which can be formed are: QT, RS, ST, TQ, TR

$$QT = \begin{bmatrix} 0 & 8 & -3 \\ 5 & 7 & -5 \\ 1 & 4 & -2 \end{bmatrix} \quad RS = \begin{bmatrix} 3 & 0 \\ 5 & 0 \\ 1 & 0 \end{bmatrix}$$

$$ST = \begin{bmatrix} 1 & 4 & -2 \end{bmatrix} \quad TQ = \begin{bmatrix} 12 & 3 \\ 5 & -2 \end{bmatrix}$$

$$TR = \begin{bmatrix} 21 & 7 \end{bmatrix}$$

2. Taking the 2-fold axis as the z-axis and the molecular plane as the (xy) plane, we have

	I	$C_2(z)$	$\sigma_{(xy)}$	i
$\Gamma_{cartesian}$	12	0	4	0

$$\Gamma_{cartesian} = 4A_g + 2A_u + 2B_g + 4B_u.$$

The table of (GX) is as follows:

3.

$\overset{\displaystyle X}{G}$	I	$S_4^2 = C_2$	S_4^1	S_4^3
I	I	C_2	S_4^1	S_4^3
$S_4^2 = C_2$	C_2	I	S_4^3	S_4^1
S_4^1	S_4^1	S_4^3	C_2	I
S_4^3	S_4^3	S_4^1	I	C_2

From this table, $I^{-1} = I$
$$C_2^{-1} = C_2$$
$$S_4^{1\,-1} = S_4^3$$
$$S_4^{3\,-1} = S_4^1$$

Then the table of $X^{-1}GX$ is

$\overset{\displaystyle G}{X}$	I	C_2	S_4^1	S_4^3
I	I	C_2	S_4^1	S_4^3
C_2	I	C_2	S_4^1	S_4^3
S_4^1	I	C_2	S_4^1	S_4^3
S_4^3	I	C_2	S_4^1	S_4^3

4. (a) $\Gamma = 2A_1 + E + 2F_2$
   (b) $\Gamma = 2A_1' + E' + E''$
   (c) $\Gamma = A_1 + A_2 + 2E$
   (d) $\Gamma = A_1 + B_1 + E$

5. The group multiplication table for C_{2v} is

	I	C_2	σ_v	σ_v'
I	I	C_2	σ_v	σ_v'
C_2	C_2	I	σ_v'	σ_v
σ_v	σ_v	σ_v'	I	C_2
σ_v'	σ_v'	σ_v	C_2	I

In Γ_a we have $\sigma_v \times C_2 = 1 \times -1 = -1 \neq \sigma_v' \ (= 1)$; thus Γ_a is not a true representation of C_{2v}.

For Γ_b, multiplication of the matrices gives:

$$
\begin{array}{c|cccc}
 & \begin{bmatrix} 1 & 0 \\ 0 & 1 \end{bmatrix} & \begin{bmatrix} -\tfrac{1}{2} & 0 \\ 0 & -\tfrac{1}{2} \end{bmatrix} & \begin{bmatrix} \tfrac{1}{2} & 0 \\ 0 & \tfrac{1}{2} \end{bmatrix} & \begin{bmatrix} -1 & 0 \\ 0 & -1 \end{bmatrix} \\[2.2em]
\hline
\begin{bmatrix} 1 & 0 \\ 0 & 1 \end{bmatrix} & \begin{bmatrix} 1 & 0 \\ 0 & 1 \end{bmatrix} & \begin{bmatrix} -\tfrac{1}{2} & 0 \\ 0 & -\tfrac{1}{2} \end{bmatrix} & \begin{bmatrix} \tfrac{1}{2} & 0 \\ 0 & \tfrac{1}{2} \end{bmatrix} & \begin{bmatrix} -1 & 0 \\ 0 & -1 \end{bmatrix} \\[2.2em]
\begin{bmatrix} -\tfrac{1}{2} & 0 \\ 0 & -\tfrac{1}{2} \end{bmatrix} & \begin{bmatrix} -\tfrac{1}{2} & 0 \\ 0 & -\tfrac{1}{2} \end{bmatrix} & \begin{bmatrix} \tfrac{1}{4} & 0 \\ 0 & \tfrac{1}{4} \end{bmatrix}^{*} & \begin{bmatrix} -\tfrac{1}{4} & 0 \\ 0 & -\tfrac{1}{4} \end{bmatrix}^{*} & \begin{bmatrix} \tfrac{1}{2} & 0 \\ 0 & \tfrac{1}{2} \end{bmatrix} \\[2.2em]
\begin{bmatrix} \tfrac{1}{2} & 0 \\ 0 & \tfrac{1}{2} \end{bmatrix} & \begin{bmatrix} \tfrac{1}{2} & 0 \\ 0 & \tfrac{1}{2} \end{bmatrix} & \begin{bmatrix} -\tfrac{1}{4} & 0 \\ 0 & -\tfrac{1}{4} \end{bmatrix}^{*} & \begin{bmatrix} \tfrac{1}{4} & 0 \\ 0 & \tfrac{1}{4} \end{bmatrix}^{*} & \begin{bmatrix} -\tfrac{1}{2} & 0 \\ 0 & -\tfrac{1}{2} \end{bmatrix} \\[2.2em]
\begin{bmatrix} -1 & 0 \\ 0 & -1 \end{bmatrix} & \begin{bmatrix} -1 & 0 \\ 0 & -1 \end{bmatrix} & \begin{bmatrix} \tfrac{1}{2} & 0 \\ 0 & \tfrac{1}{2} \end{bmatrix} & \begin{bmatrix} -\tfrac{1}{2} & 0 \\ 0 & -\tfrac{1}{2} \end{bmatrix} & \begin{bmatrix} 1 & 0 \\ 0 & 1 \end{bmatrix}
\end{array}
$$

The four asterisked product matrices do not satisfy the multiplication table. Note that the character of the representation formed by these matrices is

	I	C_2	σ_v	σ_v'
$\chi =$	2	-1	1	-2

which reduces to $\Gamma = \tfrac{1}{2}A_2 + \tfrac{3}{2}B_1$, again showing that it is not a true representation.

Chapter 6

1. Diffraction will be given for the plane with the given d-spacings at the following angles of θ, 8·88Å at 5°, 4·44Å at 10°, 3·19Å at 14°, 2·34Å at 17°, 2·25Å at 20°, 1·89Å at 24°, 1·31Å at 36°. 1·00Å does not give a diffracted beam at any of the θ values listed.

2. $d_1, 001; d_2, 101; d_3, 10\bar{1}; d_4, 100; d_5, 20\bar{1}; d_6, 202$.

3. (i) Mo, cubic $a = 4·16$; (ii) sucrose, monoclinic,
$a = 10·91, b = 8·71, c = 7·76, \beta = 103°$, (iii) SnS, orthorhombic, $a = 3·99$,
$b = 4·34, c = 11·17$; (iv) $Sm(OH)_3$, hexagonal, $a = b = 6·32, c = 3·60, \gamma = 60°$,
(v) urea, tetragonal, $a = b = 5·64, c = 4·70$.

4. (i) $SnI_4 - P$, (ii) $CdF_2 - F$, (iii) $PaPd_3 - P$, (iv) α-Sn $- F$, (v) $Y_2O_3 - I$.

5. (a) 16 (b) 4 (c) 4 (d) 8 (e) 4 (f) 3 (g) 6 (h) 1.

6. (a) Fdd2 (b) $P2_1/c$ (c) $Pna2_1$ (d) Pa3 (e) $I4_1/a$.

7. (i) Ni atoms must be at the centre of symmetry.
(ii) A atoms must be on the 2-fold axes, e.g. at $0, 0, z$ and $\frac{1}{2}, \frac{1}{2}, z$ or at $0, \frac{1}{2}, z$
and $\frac{1}{2}, 0, z$.
(iii) The X atoms must lie on both mirror planes. The Y atoms must lie on
one of the mirror planes.
(iv) AB_2 with a cell content of 2 could not have space group $Pna2_1$, because
there are no 2-fold special positions in this symmetry to accommodate the
two A atoms.

8. The point group of each molecule is given and it is also stated whether or not
the molecule has a permanent dipole.
(a) C_{2v} yes; (b) C_{2v}, yes; (c) D_{2h}, no; (d) C_{2v}, yes; (e) C_2, yes; (f) $C_s \equiv C_{1v}$, yes
(g) C_{2v}, yes; (h) C_{3v}, yes; (i) C_{4v}, yes; (j) C_{2v}, yes; (k) C_{2h}, no.

9. (a) No. A and B both belong to C_{nv} point groups, which are C_{3v} and C_{2v}
respectively.
(b) No. Cyclopropene and methylacetylene have point groups C_{2v} and C_{3v}
respectively.
(c) No. The configuration with the linear BOH systems has the point group
D_{3h}; the other configuration has the point group C_{3h}.
(d) No. The unoccupied axial and equatorial positions lead to the point groups
C_{3v} and C_{2v} respectively.
(e) Yes. The ethylenediamine ligand itself has C_{2v} symmetry, giving the point
groups D_{2h} for the trans- and C_2 for the cis-isomers.
(f) Yes. The square-based pyramid and trigonal bipyramid belong to the point
groups C_{4v} and D_{3h} respectively.

10. The point group of each molecule is given, and it is also stated whether or not
the molecule or ion is expected to be optically active.
(a) D_2, yes; (b) C_s, no; (c) C_1, yes; (d) D_{2d}, no; (e) C_{2v}, no; (f) C_{2v}, no;
(g) C_2, yes; (h) C_s, no; (i) C_{2v}, no; (j) C_2, yes.

11. (a) Yes. The *cis*-form is optically active but the *trans*-form is not.

(b) No. Both configurations have S_n axes; the planes of symmetry are equivalent to S_1. In addition, the second configuration has an S_3 axis passing through the axial CN groups and the central Ni atom (assuming the Ni $-$ C$\equiv$N system to be linear).

12. Any complex Ptabcd where a, b, c, d are all different substituents may be used as long as the plane containing them is a plane of symmetry. In this case, the complex will have C_s symmetry if the arrangement of the bonds is square-planar and C_1 if it is tetragonal-pyramidal. The second of these configurations should display optical activity. Pta$_2$bc is also suitable provided that the two identical substituents occupy *cis*-positions.

13. The calculation table and Γ_{vib} are given for each molecule or ion.

(a) H_2O_2

| | Proper | | |
	R	I	C_2	$h = 2$
n_R		4	0
$\chi_R = \pm 1 + 2\cos\theta$		3	-1
$\chi_0 = n_R \chi_R$		12	0
$\chi_{trans} = \chi_R$		3	-1
$\chi_{rot} = \pm\chi_{trans}$		3	-1
$\chi_{vib} =$		6	$+2$
$\chi_0 - \chi_{rot} - \chi_{trans}$			

$$\Gamma_{vib} = 4A + 2B$$

(b) $COCl_2$

| | | Proper | | Improper | | |
	R	I	$C_{2(z)}$	$\sigma_v(yz)$	$\sigma_v(xz)$	$h = 4$
n_R		4	2	4	2
χ_R		3	-1	1	1
χ_0		12	-2	4	2
χ_{trans}		3	-1	1	1
χ_{rot}		3	-1	-1	-1
χ_{vib}		6	0	4	2

$$\Gamma_{vib} = 3A_1 + B_1 + 2B_2$$

If the (xz) plane is taken as the molecular plane, so that the values of n_R for $\sigma_v(yz)$ and $\sigma_v(xz)$ are reversed, Γ_{vib} becomes $3A_1 + 2B_1 + B_2$. This ambiguity always occurs with molecules of C_{2v} symmetry.

(c) BrO_3^-

R	Proper			Improper	$h = 6$
	I	$2C_3$		$3\sigma_v$	
n_R	4	1		2	
χ_R	3	0		1	
χ_0	12	0		2	
χ_{trans}	3	0		1	
χ_{rot}	3	0		−1	
χ_{vib}	6	0		2	

$$\Gamma_{vib} = 3A_1 + 3E$$

Note that this is the first problem in which we have had to take g_R into account in the first two problems, there was only one symmetry operation in each class. In this case we have $3\sigma_v$, so that

$g_R\chi_{vib} =$	I	$2C_3$	$3\sigma_v$
	6	0	6

(d) p-dichlorobenzene

R	Proper				Improper			$h = 8$
	I	$C_2(z)$	$C_2(y)$	$C_2(x)$	i	$\sigma(xy)$	$\sigma(xz)$	$\sigma(yz)$
n_R	12	0	4	0	0	12	0	4
χ_R	3	−1	−1	−1	−3	1	1	1
χ_0	36	0	−4	0	0	12	0	4
χ_{trans}	3	−1	−1	−1	−3	1	1	1
χ_{rot}	3	−1	−1	−1	3	−1	−1	−1
χ_{vib}	30	2	−2	2	0	12	0	4

$\Gamma_{vib} = 6A_g + 5B_{1g} + B_{2g} + 3B_{3g} + 2A_u + 2B_{1u} + 5B_{2u} + 5B_{3u}$. The same type of ambiguity arises as in problem (b). The answer given above is correct for the labelling of planes and axes given on page 199.

(e) $XeOF_4$

		Proper		Improper		$h = 8$
R	I	$2C_4$	C_2	$2\sigma_v$	$2\sigma_d$	
n_R	6	2	2	4	2	
χ_R	3	1	−1	1	1	
χ_0	18	2	−2	4	2	
χ_{trans}	3	1	−1	1	1	
χ_{rot}	3	1	−1	−1	−1	
χ_{vib}	12	0	0	4	2	
$g_R\chi_{vib}$	12	0	0	8	4	

$$\Gamma_{vib} = 3A_1 + 2B_1 + B_2 + 3E$$

It is conventional to take the σ_v as the symmetry planes containing the Xe—F bonds and the σ_d as the symmetry planes bisecting the $F\hat{X}eF$ angles.

(f) B_2Cl_4

		Proper		Improper		$h = 8$
R	I	C_2	$2C_2$	$2S_4$	$2\sigma_d$	
n_R	6	2	0	0	4	
χ_R	3	−1	−1	−1	1	
χ_0	18	−2	0	0	4	
χ_{trans}	3	−1	−1	−1	1	
χ_{rot}	3	−1	−1	1	−1	
χ_{vib}	12	0	2	0	4	
$g_R\chi_{vib}$	12	0	4	0	8	

$$\Gamma_{vib} = 3A_1 + B_1 + 2B_2 + 3E$$

Note that although the two boron atoms lie on the S_4 axis, they are not invariant to the operation because they are exchanged by the reflection which follows the rotation through 90°. Also note that in addition to the C_2 which is equivalent to S_4^2, there are two other 2-fold axes labelled $2C_2'$. They are distinguished from C_2 because they do not leave the same atoms invariant.

(g) MoF_6

	Proper					Improper					$h = 48$
R	I	$8C_3$	$3C_2$	$6C_4$	$6C_2'$	i	$8S_6$	$3\sigma_h$	$6S_4$	$6\sigma_d$	
n_R	7	1	3	3	1	1	1	5	1	3	
χ_R	3	0	-1	1	-1	-3	0	1	-1	1	
χ_0	21	0	-3	3	-1	-3	0	5	-1	3	
χ_{trans}	3	0	-1	1	-1	-3	0	1	-1	1	
χ_{rot}	3	0	-1	1	-1	3	0	-1	1	-1	
χ_{vib}	15	0	-1	1	1	-3	0	5	-1	3	
$g_R\chi_{vib}$	15	0	-3	6	6	-3	0	15	-6	18	

$$\Gamma_{vib} = A_{1g} + E_g + T_{2g} + 2T_{1u} + T_{2u}$$

In this molecule, there are three 4-fold axes, which lie along the bonds directed towards atoms which are *trans* to each other. These axes generate $6C_4$ and $3C_2$ ($\equiv C_4^2$); they are also S_4 axes. Note that the same atoms are not invariant to C_4 and S_4 since, as in the previous example, the end atoms are interchanged by the reflection part of the S_4 operation. The $3\sigma_h$ each contain four bonds directed to the corners of a plane square. The $3C_2'$ are formed by the intersection of the σ_d with the σ_h; the σ_d contain two collinear bonds and bisect a pair of inter-bond angles.

(h) $AuCl_4^-$

	Proper					Improper					$h = 16$
R	I	$2C_4$	C_2	$2C_2'$	$2C_2''$	i	$2S_4$	σ_h	$2\sigma_v$	$2\sigma_d$	
n_R	5	1	1	3	1	1	1	5	3	1	
χ_R	3	1	-1	-1	-1	-3	-1	1	1	1	
χ_0	15	1	-1	-3	-1	-3	-1	5	3	1	
χ_{trans}	3	1	-1	-1	-1	-3	-1	1	1	1	
χ_{rot}	3	1	-1	-1	-1	3	1	-1	-1	-1	
χ_{vib}	9	-1	1	-1	1	-3	-1	5	3	1	
$g_R\chi_{vib}$	9	-2	1	-2	2	-3	-2	5	6	2	

$$\Gamma_{vib} = A_{1g} + B_{1g} + B_{2g} + A_{2u} + B_{2u} + 2E_u$$

In this molecule there are three different types of C_2 axis; $C_2 \equiv C_4^2$, C_2' formed by the intersection of σ_v with the molecular plane σ_h and C_2'' formed by the intersection of σ_d with σ_h. The σ_v contain bonds and the σ_d bisect inter-bond angles. The C_4 is also an S_4 and the $C_2 \equiv S_4^2$ is thus identical with the $C_2 \equiv C_4^2$.

(i) C_2H_6

R	Proper			Improper		
	I	$2C_3$	$3C_2$	i	$2S_6$	$3\sigma_d$
n_R	8	2	0	0	0	4
χ_R	3	0	-1	-3	0	1
χ_0	24	0	0	0	0	4
χ_{trans}	3	0	-1	-3	0	1
χ_{rot}	3	0	-1	3	0	-1
χ_{vib}	18	0	2	0	0	4
$g_R\chi_{vib}$	18	0	6	0	0	12

$h = 12$

$$\Gamma_{vib} = 3A_{1g} + A_{1u} + 2A_{2u} + 3E_g + 3E_u$$

(j) CCl_4

R	Proper			Improper	
	I	$8C_3$	$3C_2$	$6S_4$	$6\sigma_d$
n_R	5	2	1	1	3
χ_R	3	0	-1	-1	1
χ_0	15	0	-1	-1	3
χ_{trans}	3	0	-1	-1	1
χ_{rot}	3	0	-1	1	-1
χ_{vib}	9	0	1	-1	3
$g_R\chi_{vib}$	9	0	3	-6	18

$h = 24$

$$\Gamma_{vib} = A_1 + E + 2T_2$$

The $3C_2$ are generated by the S_4 ($C_2 \equiv S_4^2$).

14. (a) C_{3v}

R	I	$2C_3$	$3\sigma_v$
$\chi_\mu = \pm 1 + 2\cos\theta$	3	0	1
$2\cos\theta$	2	-1	2
$\chi_\alpha = (2\cos\theta)\chi_\mu$	6	0	2

$h = 6$

$$\Gamma_\mu = A_1 + E; \quad \Gamma_a = 2A_1 + 2E$$

Thus vibrations of classes A_1 and E are active in both the infra-red and Raman effect.

(b) D_{2d}

R	I	C_2	$2C_2'$	$2S_4$	$2\sigma_d$	$h = 8$
χ_μ	3	−1	−1	−1	1	
$2\cos\theta$	2	−2	−2	0	2	
χ_α	6	2	2	0	2	

$\Gamma_\mu = B_2 + E$; $\Gamma_\alpha = A_1 + B_1 + B_2 + E$

(c) D_2

R	I	$C_{2(z)}$	$C_{2(y)}$	$C_{2(x)}$	$h = 4$
χ_μ	3	−1	−1	−1	
$2\cos\theta$	2	−2	−2	−2	
χ_α	6	2	2	2	

$\Gamma_\mu = B_1 + B_2 + B_3$; $\Gamma_\alpha = 3A_1 + B_1 + B_2 + B_3$

(d) D_{6h}

R	I	$2C_6$	$2C_3$	C_2	$3C_2'$	$3C_2''$	i	$2S_3$	$2S_6$	σ_h	$3\sigma_d$	$3\sigma_v$	$h = 24$
χ_μ	3	2	0	−1	−1	−1	−3	−2	0	1	1	1	
$2\cos\theta$	2	1	−1	−2	−2	−2	−2	−1	1	2	2	2	
χ_α	6	2	0	2	2	2	6	2	0	2	2	2	

$\Gamma_\mu = A_{2u} + E_{1u}$; $\Gamma_\alpha = 2A_{1g} + E_{1g} + E_{2g}$.

15. Consider the transformations of a general point (x,y,z) under the symmetry operations of C_{2v}

$(x,y,z) \xrightarrow{\ I\ } (x,y,z)$;   transformation matrix $\begin{bmatrix} 1 & 0 & 0 \\ 0 & 1 & 0 \\ 0 & 0 & 1 \end{bmatrix}$

$(x,y,z) \xrightarrow{\ C_2(z)\ } (-x,-y,z)$ $\begin{bmatrix} -1 & 0 & 0 \\ 0 & -1 & 0 \\ 0 & 0 & 1 \end{bmatrix}$

$(x,y,z) \xrightarrow{\ \sigma_v(yz)\ } (-x, y, z)$ $\begin{bmatrix} -1 & 0 & 0 \\ 0 & 1 & 0 \\ 0 & 0 & 1 \end{bmatrix}$

$(x,y,z) \xrightarrow{\ \sigma_v(xz)\ } (x,-y,z)$ $\begin{bmatrix} 1 & 0 & 0 \\ 0 & -1 & 0 \\ 0 & 0 & 1 \end{bmatrix}$

Then we may list these transformations as follows:

	I	$C_2(z)$	$\sigma_v(yz)$	$\sigma_v(xz)$
x	1	-1	-1	1
y	1	-1	1	-1
z	1	1	1	1

Comparing these with the character table of C_{2v}, we see that z has the same character as A_1, x as B_1 and y as B_2

The initial vibrational state is the totally symmetric A_1 state.

The character of the direct product ($\phi_i M \phi_f$) has to be the character of the totally symmetric representation in order for the transition to occur. From the character table we have

$$\chi(A_1\ zA_1) = \chi(A_1); \quad \chi(A_1 xB_1) = \chi(A_1); \quad \chi(A_1 yB_2) = \chi(A_1)$$

For the Raman selection rules we need to form the characters of the direct products x^2, y^2, z^2, xy, xz and yz

	I	$C_2(z)$	$\sigma_v(yz)$	$\sigma_v(xz)$	
x^2, y^2, z^2	1	1	1	1	A_1
xz	1	-1	-1	1	B_1
yz	1	-1	1	-1	B_2
xy	1	1	-1	-1	A_2

Combinations whose character is the character of A_1 are therefore:

ϕ_i	α	ϕ_f	
A_1	x^2, y^2, z^2	A_1	
A_1	xz	B_1	
A_1	yz	B_2	
A_1	xy	A_2	confirming the selection rules given.

16. $\phi_i(\phi_2)B_1$
            $\phi_f(\phi_5)A_1$ $(\phi_6)B_2$

$\phi_i \phi_f$	B_1	A_2
Component		
$x(B_1)$	A_1	B_2
$y(B_2)$	A_2	A_1
$z(A_1)$	B_1	A_2

Since the combinations (ϕ_2 x ϕ_5) and (ϕ_2 y ϕ_6) have the character of A_1, the transitions from ϕ_2 to ϕ_5 in the x-direction and from ϕ_2 to ϕ_6 in the y-direction are permitted.

Chapter 7

1. Best to draw diagrams of the orbitals and operate the elements of the group on them.

2. The p- and d-orbitals will be split into sets of symmetry:

	symmetry of field	p_x	p_y	p_z	d_{z^2}	$d_{x^2-y^2}$	d_{xy}	d_{xz}	d_{yz}
a	D_{4h}	E_u		A_{2u}	A_{1g}	B_{1g}	B_{2g}	E_g	
b	D_{4d}	E_1		B_2	A_1	E_2		E_3	
c	D_3	E		A_2	A_1	E		E	
d	D_{3h}	E'		A_2''	A_1'	E'		E''	
e	D_{2d}	E_1		B_2	A_1	B_1	B_2	E	
f	C_{4v}	E		A_1	A_1	B_1	B_2	E	
g	C_{2v}	B_1	B_2	A_1	A_1	A_1	A_2	B_2	B_1

3. (a) $\Psi_{SF_6} = A_{1g} + T_{1u} + E_g$
   i.e. s, p_x, p_y, p_z, $d_{x^2-y^2}$ and d_{z^2} orbitals

   (b) $\Psi_{XeF_4} = A_{1g} + E_u + B_{1g}$
   i.e. s, p_x, p_y and $d_{x^2-y^2}$ orbitals

4. The orbitals would be split into 3 sets

(i) f_{xyz} with symmetry A_{2u}
(ii) The triply degenerate set f_{x^3}, f_{y^3}, f_{z^3} with symmetry T_{1u}
(iii) The triply degenerate set $f_{x(y^2-z^2)}$, $f_{y(z^2-x^2)}$, $f_{z(x^2-y^2)}$ with symmetry T_{2u}

5. Labelling the carbon atoms $1\!\!-\!\!\overset{2}{-\!\!-}\!\!-3$, we proceed as follows:

Step 1   The allyl system has the point group C_{2v} and its rotation sub-group is C_2

Step 2   The transformation table for the orbitals under the operations of the point group C_2 is

	I	C_2
Φ_1	Φ_1	Φ_3
Φ_2	Φ_2	Φ_2
Φ_3	Φ_3	Φ_1
Number of unshifted orbitals	3	1

Using Table 7.10 we have $a_{(A)} = \frac{1}{2}[1.1.3 + 1.1.1] = 2$
$a_{(B)} = \frac{1}{2}[1.1.3 - 1.1.1] = 1$

Thus we need two symmetry-adapted orbitals of class A and one of class B.

Step 3   Using Φ_1 as a generating function we have

$$\phi_1(A) = I\Phi_1.1 + C_2\Phi_1.1$$
$$= \Phi_1 + \Phi_3$$
$$= \frac{1}{\sqrt{2}}(\Phi_1 + \Phi_3) \text{ on normalisation}$$

The same symmetry-adapted orbital would be obtained using Φ_3 as the generating function.

Using Φ_2 as a generating function we have

$$\phi_2(A) = I\Phi_2.1 + C_2\Phi_2.1$$
$$= \Phi_2 + \Phi_2$$
$$= \Phi_2 \text{ on normalisation}$$

Using Φ_1 as a generating function for an orbital of class B we have

$$\phi_3(B) = I\Phi_1.1 + C_2\Phi_1.(-1)$$
$$= \Phi_1 - \Phi_3$$
$$= \frac{1}{\sqrt{2}}[\Phi_1 - \Phi_3] \text{ on normalisation.}$$

If Φ_3 is used instead of Φ_1, the signs of the coefficients will be reversed. It is not possible to generate an orbital of class B from Φ_2.

Now
$$H_{11} = \int \phi_1 \mathcal{H}\phi_1 \, d\tau$$
$$= \frac{1}{\sqrt{2}} \times \frac{1}{\sqrt{2}} \int [\Phi_1 + \Phi_3] \mathcal{H}[\Phi_1 + \Phi_3] \, d\tau$$
$$= \frac{1}{2}\int [\Phi_1 \mathcal{H}\Phi_1 + \Phi_3\mathcal{H}\Phi_3 + 2\Phi_1\mathcal{H}\Phi_3] d\tau$$
$$= \frac{1}{2}[\alpha + \alpha + 0]$$
$$= \alpha.$$

$$H_{22} = \int \phi_2 \mathcal{H}\phi_2 \, d\tau$$
$$= \int \Phi_2 \mathcal{H}\Phi_2 \, d\tau$$
$$= \alpha.$$

$$H_{12} = \int \phi_1 \mathcal{H}\phi_2 \, d\tau = \frac{1}{\sqrt{2}} \int [\Phi_1 + \Phi_3] \mathcal{H}\Phi_2 \, d\tau$$
$$= \frac{1}{\sqrt{2}}[2\beta] = \sqrt{2}\beta$$

Thus the secular determinant for the energy levels corresponding to the orbitals of class A is

$$\begin{vmatrix} \alpha - E & \sqrt{2}\beta \\ \sqrt{2}\beta & \alpha - E \end{vmatrix} = 0 \qquad \text{whose solutions are } E = \alpha \pm \sqrt{2}\beta$$

For the orbital of class B,

$$H_{33} = \int \phi_3 \mathcal{H} \phi_3 \, d\tau = \frac{1}{\sqrt{2}} \times \frac{1}{\sqrt{2}} \int [(\Phi_1 - \Phi_3)\mathcal{H}(\Phi_1 - \Phi_3)] \, d\tau$$

$$= \frac{1}{2}[2\alpha]$$

$$= \alpha$$

Thus for the orbital of class B the determinant is simply

$$|\alpha - E| = 0 \text{ so } E = \alpha.$$

Since H_{33} is the only term, the symmetry-adapted orbital is the true molecular orbital.

For the orbitals of class A, consider the possibility

$$\psi_1 = \frac{1}{(1+x^2)^{\frac{1}{2}}}[\phi_1 + x\phi_2], \text{ with } E = \alpha + \sqrt{2}\beta$$

Then $\quad \alpha + \sqrt{2}\beta = \int \psi_1 \mathcal{H} \psi_1 \, d\tau$

$$= \frac{1}{1+x^2} \int (\phi_1 + x\phi_2)\mathcal{H}(\phi_1 + x\phi_2) \, d\tau$$

$$= \frac{1}{1+x^2} \int [\phi_1 \mathcal{H} \phi_1 + 2x\phi_1 \mathcal{H} \phi_2 + x^2 \phi_2 \mathcal{H} \phi_2] \, d\tau$$

$$= \frac{1}{1+x^2} [\alpha + 2\sqrt{2}x\beta + x^2 \alpha]$$

$\therefore \quad \sqrt{2}(1+x^2) = 2\sqrt{2}x$

$\therefore \quad x = +1$

$\therefore \quad \psi_1 = \left(\frac{1}{1+1^2}\right)^{\frac{1}{2}} \left[\frac{1}{\sqrt{2}}(\Phi_1 + \Phi_3) + \Phi_2\right]$

$$= \frac{1}{\sqrt{2}} \left[\frac{1}{\sqrt{2}}(\Phi_1 + \Phi_3) + \Phi_2\right]$$

$$= \frac{1}{2}[\Phi_1 + \sqrt{2}\,\Phi_2 + \Phi_3]$$

For the other orbital of class A, with $E = \alpha - \sqrt{2}\beta$ we have

$$\psi_2 = \left(\frac{1}{1+x^2}\right)^{\frac{1}{2}} [\phi_1 - x\phi_2]$$

$$\alpha - \sqrt{2}\beta = \int \psi_2 \mathcal{H} \psi_2 d\tau$$

$$= \left(\frac{1}{1+x^2}\right) \int [\phi_1 \mathcal{H}\phi_1 - 2x\phi_1 \mathcal{H}\phi_2 + x^2 \phi_2 \mathcal{H}\phi_2] d\tau$$

$$= \left(\frac{1}{1+x^2}\right) [\alpha - 2\sqrt{2}\beta x + x^2 \alpha]$$

$$\therefore \qquad -\sqrt{2}(1+x^2) = -2\sqrt{2}x$$

and $\qquad\qquad x = 1$ as before, giving

$$\psi_3 = \frac{1}{\sqrt{2}} \left[\frac{1}{\sqrt{2}}(\Phi_1 + \Phi_3) - \Phi_2 \right]$$

$$= \frac{1}{2} [\Phi_1 - \sqrt{2}\Phi_2 + \Phi_3]$$

6. Cyclopropene with the equilateral triangular shape has the point group D_{3h} and we can thus classify the orbitals according to the irreducible representation of the point group C_3. The transformation table for the orbitals is

	I	C_3^1	C_3^2
Φ_1	1	2	3
Φ_2	2	3	1
Φ_3	3	1	2

As with benzene we need to construct one orbital of each irreducible representation of the C_n point group. Using Φ_1 as the generating function, we have the un-normalised orbitals

$$\phi_1(A) = (\Phi_1 + \Phi_2 + \Phi_3)$$

$$\phi_2'(E) = \Phi_1 + \epsilon\Phi_2 + \epsilon^*\Phi_3$$

$$\phi_3'(E) = \Phi_1 + \epsilon^*\Phi_2 + \epsilon\Phi_3$$

In this case $\quad \epsilon = \cos\frac{2\pi}{3} + i\sin\frac{2\pi}{3}$

$$\epsilon^* = \cos\frac{2\pi}{3} - i\sin\frac{2\pi}{3}$$

$$\cos\frac{2\pi}{3} = -\frac{1}{2}, \; \sin\frac{2\pi}{3} = \frac{\sqrt{3}}{2}$$

Thus $\epsilon + \epsilon^* = -1, \epsilon - \epsilon^* = i\sqrt{3}$

Forming $\phi_2 = \phi_2' + \phi_3'$ we obtain

$$\phi_2 = 2\Phi_1 - \Phi_2 - \Phi_3$$

and from $i\phi_3 = \phi_2' - \phi_3'$ we have

$$\phi_3 = \sqrt{3}(\Phi_2 - \Phi_3)$$

Normalising these orbitals we have

$$\phi_1 = \frac{1}{\sqrt{3}}(\Phi_1 + \Phi_2 + \Phi_3)$$

$$\phi_2 = \frac{1}{\sqrt{6}}(2\Phi_1 - \Phi_2 - \Phi_3)$$

$$\phi_3 = \frac{1}{\sqrt{2}}(\Phi_2 - \Phi_3)$$

$$E_1 = \int \phi_1 \mathcal{H} \phi_1 d\tau = \frac{1}{3}\int [\Phi_1 + \Phi_2 + \Phi_3]\mathcal{H}[\Phi_1 + \Phi_2 + \Phi_3]d\tau$$

$$= \frac{1}{3}[3\alpha + 6\beta]$$

$$= \alpha + 2\beta$$

$$E_2 = \int \phi_2 \mathcal{H} \phi_2 d\tau = \frac{1}{6}\int [2\Phi_1 - \Phi_2 - \Phi_3]\mathcal{H}[2\Phi_1 - \Phi_2 - \Phi_3]d\tau$$

$$= \frac{1}{6}[6\alpha - 6\beta] = \alpha - \beta$$

$$E_3 = \int \phi_3 \mathcal{H} \phi_3 d\tau = \frac{1}{2}\int [\Phi_2 - \Phi_3]\mathcal{H}[\Phi_2 - \Phi_3]d\tau$$

$$= \frac{1}{2}[2\alpha - 2\beta] = \alpha - \beta.$$

The true molecular orbital ψ_1 is equal to ϕ_1 since it is the only orbital of class A. Also, $H_{23} = H_{32} = \int \phi_2 \mathcal{H} \phi_3 d\tau$

$$= \frac{1}{\sqrt{12}}\int [2\Phi_1 - \Phi_2 - \Phi_3]\mathcal{H}[\Phi_2 - \Phi_3]d\tau$$

$$= 0$$

and since $H_{23} = H_{32} = 0$, ϕ_2 and ϕ_3 are the true molecular orbitals ψ_2 and ψ_3.

Cyclopropene with the isosceles triangular shape is the same, from the symmetry aspect, as the allyl system. The calculation is exactly similar to that for allyl except that, as atoms 1 and 3 are now adjacent atoms of the cyclic system, $H_{13} = H_{31} = \beta$.

Thus for $\phi_1 = \dfrac{1}{\sqrt{2}}(\Phi_1 + \Phi_3),$ $H_{11} = \alpha + \beta$

$\phi_2 = \Phi_2,$ $H_{22} = \alpha$

$$H_{12} = \int \phi_1 \mathcal{H} \phi_2 d\tau = \sqrt{2}\beta \quad \therefore \quad \begin{bmatrix} \alpha + \beta - E & \sqrt{2}\beta \\ \sqrt{2}\beta & \alpha - E \end{bmatrix} = 0$$

$$E = \alpha + 2\beta \text{ or } \alpha - \beta.$$

$\phi_3 = \dfrac{1}{\sqrt{2}}(\Phi_1 - \Phi_3),$ $H_{33} = \alpha - \beta$

Since H_{12} is not zero, ϕ_1 and ϕ_2 are not true molecular orbitals

$$\psi_1 = \left(\frac{1}{1 + x^2}\right)^{\frac{1}{2}} [\phi_1 + x\phi_2]$$

$$\begin{aligned} E_1 = \alpha + 2\beta &= \int \psi_1 \mathcal{H} \psi_1 d\tau \\ &= \left(\frac{1}{1 + x^2}\right) \int [\phi_1 + x\phi_2] \mathcal{H} [\phi_1 + x\phi_2] d\tau \\ &= \frac{1}{1 + x^2} \int [\phi_1 \mathcal{H}\phi_1 + 2x\phi_1 \mathcal{H}\phi_2 + x^2 \phi_2 \mathcal{H}\phi_2] d\tau \\ &= \frac{1}{1 + x^2} [(\alpha + \beta) + 2\sqrt{2}\beta x + x^2 \alpha] \end{aligned}$$

$$\therefore \ (\alpha + 2\beta)(1 + x^2) = \alpha(1 + x^2) + \beta(1 + 2\sqrt{2}x) \text{ from which } x = \frac{1}{\sqrt{2}}$$

Thus $\psi_1 = \dfrac{1}{\sqrt{3/2}} \left[\dfrac{1}{\sqrt{2}}(\Phi_1 + \Phi_3) + \dfrac{1}{\sqrt{2}}\phi_2 \right]$

$$= \frac{1}{\sqrt{3}}[\Phi_1 + \Phi_2 + \Phi_3]$$

$$\psi_2 = \left(\frac{1}{1 + x^2}\right)^{\frac{1}{2}} [\phi_1 - x\phi_2]$$

$$\begin{aligned} E_2 = \alpha - \beta &= \int \psi_2 \mathcal{H} \psi_2 d\tau \\ &= \left(\frac{1}{1 + x^2}\right) \int [\phi_1 - x\phi_2] \mathcal{H} [\phi_1 - x\phi_2] d\tau \\ &= \frac{1}{1 + x^2} \int (\phi_1 \mathcal{H}\phi_1 - 2x\phi_1 \mathcal{H}\phi_2 + x^2 \phi_2 \mathcal{H}\phi_2) d\tau \end{aligned}$$

$$= \frac{1}{1+x^2} [\alpha + \beta - 2\sqrt{2}\beta x + x^2 \alpha]$$

$$-\beta(1+x^2) = \beta - 2\sqrt{2}\beta x$$

$$\therefore \qquad x^2 - 2\sqrt{2}x + 2 = 0$$

$$\therefore \qquad x = \sqrt{2}$$

$$\therefore \qquad \psi_2 = \frac{1}{\sqrt{6}} [\Phi_1 - 2\Phi_2 + \Phi_3]$$

$$\psi_3 = \phi_3 = \frac{1}{\sqrt{2}} (\Phi_2 - \Phi_3)$$

So the change in shape of the cyclopropene molecule does not affect either the energy levels or the molecular orbital coefficients at this level of approximation.

7. The character $\chi(R)$ of the p-orbitals under the symmetry operations of the full point group of benzene is:

	I	$2C_6^6$	$2C_3$	C_2	$3C_2'$	$3C_2''$	i	$2S_3$	$2S_6$	σ_h	$3\sigma_d$	$3\sigma_v$
$\chi(R)$	6	0	0	0	−2	0	0	0	0	−6	0	2
$g_R \chi(R)$	6	0	0	0	−6	0	0	0	0	−6	0	6

(Remember that the operations C_2' and σ_h turn the orbital upside down, so that it makes a negative contribution to the character.)

The order h of the D_{6h} point group is 24, and from the relevant character table we have

$$\chi(R) = \chi(A_{2u}) + \chi(B_{2g}) + \chi(E_{1g}) + \chi(E_{2u})$$

Appendix II

(a) POINT GROUPS WITH NO PRINCIPAL AXIS

C_1	I
A	1

C_s	I	σ_h		
A'	1	1	R_z	x, y, x^2, y^2, z^2, xy
A''	1	-1	R_x, R_y	z, yz, xz

C_i	I	i		
A_g	1	1	R_x, R_y, R_z	$x^2, y^2, z^2, xy, xz, yz$
A_u	1	-1		x, y, z

(b) C_n POINT GROUPS

C_2 See Table 7.10 p. 173

C_3	I	C_3	C_3^2		$\epsilon = \exp(2\pi i/3)$
A	1	1	1	R_z	$z, x^2 + y^2, z^2$
E	$\begin{cases}1 \\ 1\end{cases}$	$\begin{matrix}\epsilon \\ \epsilon^*\end{matrix}$	$\begin{matrix}\epsilon^* \\ \epsilon\end{matrix}$	(R_x, R_y)	$(x, y)\ (x^2 - y^2, xy)(yz, xz)$

C_4	I	C_4	C_2	C_4^3		
A	1	1	1	1	R_z	$z, x^3 + y^2, z^2$
B	1	-1	1	-1		$x^2 - y^2, xy$
E	$\begin{cases}1 \\ 1\end{cases}$	$\begin{matrix}i \\ -i\end{matrix}$	$\begin{matrix}-1 \\ -1\end{matrix}$	$\begin{matrix}-i \\ i\end{matrix}$	(R_x, R_y)	$(x, y)(yz, xz)$

C_6 See Table 7.11 p. 179

(c) D_n POINT GROUPS

D_2	I	$C_2(z)$	$C_2(y)$	$C_2(x)$		
A	1	1	1	1		x^2, y^2, z^2
B_1	1	1	-1	-1	R_z	z, xy
B_2	1	-1	1	-1	R_y	y, xz
B_3	1	-1	-1	1	R_x	x, yz

D_3	I	$2C_3$	$3C_2$		
A_1	1	1	1		$x^2+y^2,\ z^2$
A_2	1	1	-1	R_z	z
E	2	-1	0	(R_x, R_y)	$(x, y)(x^2-y^2, xy)(xz, yz)$

D_4	I	$2C_4$	$C_2(=C_4^2)$	$2C_2'$	$2C_2''$		
A_1	1	1	1	1	1		$x^2+y^2,\ z^2$
A_2	1	1	1	-1	-1	R_z	z
B_1	1	-1	1	1	-1		x^2-y^2
B_2	1	-1	1	-1	1		xy
E	2	0	-2	0	0	(R_x, R_y)	$(x, y)(xz, yz)$

(d) S_n POINT GROUPS

S_4	I	S_4^1	C_2	S_4^3		
A	1	1	1	1	R_z	$x^2+y^2,\ z^2$
B	1	-1	1	-1		$z,\ x^2-y^2,\ xy$
E	$\begin{cases}1\\1\end{cases}$	$\begin{matrix}i\\-i\end{matrix}$	$\begin{matrix}-1\\-1\end{matrix}$	$\left.\begin{matrix}-i\\i\end{matrix}\right\}$	(R_x, R_y)	$(xz, yz)\ (x, y)$

S_6	I	C_3	C_3^2	i	S_6^5	S_6		$\epsilon = \exp(2\pi i/3)$
A_g	1	1	1	1	1	1	R_z	$x^2+y^2,\ z^2$
E_g	$\begin{cases}1\\1\end{cases}$	$\begin{matrix}\epsilon\\\epsilon^*\end{matrix}$	$\begin{matrix}\epsilon^*\\\epsilon\end{matrix}$	$\begin{matrix}1\\1\end{matrix}$	$\begin{matrix}\epsilon\\\epsilon^*\end{matrix}$	$\left.\begin{matrix}\epsilon^*\\\epsilon\end{matrix}\right\}$	(R_x, R_y)	$\begin{matrix}(x^2-y^2, xy)\\(xz, yz)\end{matrix}$
A_u	1	1	1	-1	-1	-1		z
E_u	$\begin{cases}1\\1\\1\end{cases}$	$\begin{matrix}\epsilon\\\epsilon^*\end{matrix}$	$\begin{matrix}\epsilon^*\\\epsilon\end{matrix}$	$\begin{matrix}-1\\-1\end{matrix}$	$\begin{matrix}-\epsilon\\-\epsilon^*\end{matrix}$	$\left.\begin{matrix}-\epsilon^*\\-\epsilon\end{matrix}\right\}$		(x, y)

S_8	I	S_8	C_4	S_8^3	C_2	S_8^5	C_4^3	S_8^7		$\epsilon = \exp(2\pi i/8)$
A	1	1	1	1	1	1	1	1	R_z	$x^2+y^2,\ z^2$
B	1	-1	1	-1	1	-1	1	-1		z
E_1	$\begin{cases}1\\1\end{cases}$	$\begin{matrix}\epsilon\\\epsilon^*\end{matrix}$	$\begin{matrix}i\\-i\end{matrix}$	$\begin{matrix}-\epsilon^*\\-\epsilon\end{matrix}$	$\begin{matrix}-1\\-1\end{matrix}$	$\begin{matrix}-\epsilon\\-\epsilon^*\end{matrix}$	$\begin{matrix}-i\\i\end{matrix}$	$\left.\begin{matrix}\epsilon^*\\\epsilon\end{matrix}\right\}$	(R_x, R_y)	(x, y)
E_2	$\begin{cases}1\\1\end{cases}$	$\begin{matrix}i\\-i\end{matrix}$	$\begin{matrix}-1\\-1\end{matrix}$	$\begin{matrix}-i\\i\end{matrix}$	$\begin{matrix}1\\1\end{matrix}$	$\begin{matrix}i\\-i\end{matrix}$	$\begin{matrix}-1\\-1\end{matrix}$	$\left.\begin{matrix}-i\\i\end{matrix}\right\}$		(x^2-y^2, xy)
E_3	$\begin{cases}1\\1\end{cases}$	$\begin{matrix}-\epsilon^*\\-\epsilon\end{matrix}$	$\begin{matrix}-i\\i\end{matrix}$	$\begin{matrix}\epsilon\\\epsilon^*\end{matrix}$	$\begin{matrix}-1\\-1\end{matrix}$	$\begin{matrix}\epsilon^*\\\epsilon\end{matrix}$	$\begin{matrix}i\\-i\end{matrix}$	$\left.\begin{matrix}-\epsilon\\-\epsilon^*\end{matrix}\right\}$		(xz, yz)

(e) C_{nv} POINT GROUPS

C_{2v}: See Table 7.1 p. 154
C_{3v} : See Table 7.2 p. 154

C_{4v}	I	$2C_4$	C_2	$2\sigma_v$	$2\sigma_d$		
A_1	1	1	1	1	1		$z,\ z^2$
A_2	1	1	1	-1	-1	R_z	
B_1	1	-1	1	1	-1		x^2-y^2
B_2	1	-1	1	-1	0		xy
E	2	0	-2	0	0	(R_x, R_y)	$(x, y),\ (xz, yz)$

C_{5v}	I	$2C_5$	$2C_5^2$	$5\sigma_v$		
A_1	1	1	1	1		z, z^2
A_2	1	1	1	-1	R_z	
E_1	2	$2\cos72°$	$2\cos144°$	0	(R_x, R_y)	$(x, y)\,(yz, xz)$
E_2	2	$2\cos144°$	$2\cos72°$	0		$(x^2 - y^2, xy)$

C_{6v}	I	$2C_6$	$2C_3$	C_2	$3\sigma_v$	$3\sigma_d$		
A_1	1	1	1	1	1	1		z, z^2
A_2	1	1	1	1	-1	-1	R_z	
B_1	1	-1	1	-1	1	-1		
B_2	1	-1	1	-1	-1	1		
E_1	2	1	-1	-2	0	0	(R_x, R_y)	$(x, y)(xz, yz)$
E_2	2	-1	-1	2	0	0		$(x^2 - y^2, xy)$

$C_{\infty v}$	I	$2C_\infty^\Phi$	$\ldots$	$\infty\sigma_v$		
$A_1 \equiv \Sigma^+$	1	1	$\ldots$	1		$z,\ x^2 + y^2,\ z^2$
$A_2 \equiv \Sigma^-$	1	1	$\ldots$	-1	R_z	
$E_1 \equiv \Pi$	2	$2\cos\Phi$	$\ldots$	0	(R_x, R_y)	$(x, y)(xz, yz)$
$E_2 \equiv \Delta$	2	$2\cos2\Phi$	$\ldots$	0		$(x^2 - y^2, xy)$
$E_3 \equiv \Phi$	2	$2\cos3\Phi$	$\ldots$	0		
$\ldots$	$\ldots$	$\ldots$	$\ldots$	$\ldots$		

(f) C_{nh} POINT GROUPS (C_{2h} see Chapter 6 page 144)

C_{3h}	I	C_3	C_3^2	σ_h	S_3	S_5^3		$\epsilon = \exp(2\pi i/3)$
A'	1	1	1	1	1	1	R_z	$x^2 + y^2, z^2$
E'	$\begin{cases}1\\1\end{cases}$	$\begin{matrix}\epsilon\\\epsilon^*\end{matrix}$	$\begin{matrix}\epsilon^*\\\epsilon\end{matrix}$	$\begin{matrix}1\\1\end{matrix}$	$\begin{matrix}\epsilon\\\epsilon^*\end{matrix}$	$\begin{matrix}\epsilon^*\\\epsilon\end{matrix}$		$(x, y)(x^2 - y^2, xy)$
A''	1	1	1	-1	-1	-1		z
E''	$\begin{cases}1\\1\end{cases}$	$\begin{matrix}\epsilon\\\epsilon^*\end{matrix}$	$\begin{matrix}\epsilon^*\\\epsilon\end{matrix}$	$\begin{matrix}-1\\-1\end{matrix}$	$\begin{matrix}-\epsilon\\-\epsilon^*\end{matrix}$	$\begin{matrix}-\epsilon^*\\-\epsilon\end{matrix}$	(R_x, R_y)	(xz, yz)

(g) D_{nh} POINT GROUPS

D_{2h}	I	$C_2(z)$	$C_2(y)$	$C_2(x)$	i	$\sigma(xy)$	$\sigma(xz)$	$\sigma(yz)$		
A_g	1	1	1	1	1	1	1	1		x^2, y^2, z^2
B_{1g}	1	1	-1	-1	1	1	-1	-1	R_z	xy
B_{2g}	1	-1	1	-1	1	-1	1	-1	R_y	xz
B_{3g}	1	-1	-1	1	1	-1	-1	1	R_x	yz
A_u	1	1	1	1	-1	-1	-1	-1		
B_{1u}	1	1	-1	-1	-1	-1	1	1		z
B_{2u}	1	-1	1	-1	-1	1	-1	1		y
B_{3u}	1	-1	-1	1	-1	1	1	-1		x

D_{3h} See Table 7.4 p. 158
D_{4h} See Table 7.8 p. 168

D_{5h}	I	$2C_5$	$2C_5^2$	$5C_2$	σ_h	$2S_5$	$2S_5^3$	$5\sigma_v$		
A_1'	1	1	1	1	1	1	1	1		$x^2+y^2,\ z^2$
A_2'	1	1	1	-1	1	1	1	-1	R_z	
E_1'	2	$2\cos72°$	$2\cos144°$	0	2	$2\cos72°$	$2\cos144°$	0		(x, y)
E_2'	2	$2\cos144°$	$2\cos72°$	0	2	$2\cos144°$	$2\cos72°$	0		(x^2-y^2, xy)
A_1''	1	1	1	1	-1	-1	-1	-1		
A_2''	1	1	1	-1	-1	-1	-1	1		z
E_1''	2	$2\cos72°$	$2\cos144°$	0	-2	$-2\cos72°$	$-2\cos144°$	0	(R_x, R_y)	(zx, yz)
E_2''	2	$2\cos144°$	$2\cos72°$	0	-2	$-2\cos144°$	$-2\cos72°$	0		

D_{6h}	I	$2C_6$	$2C_3$	C_2	$3C_2'$	$3C_2''$	i	$2S_3$	$2S_6$	σ_h	$3\sigma_d$	$3\sigma_v$		
A_{1g}	1	1	1	1	1	1	1	1	1	1	1	1		$x^2+y^2,\ z^2$
A_{2g}	1	1	1	1	-1	-1	1	1	1	1	-1	-1	R_z	
B_{1g}	1	-1	1	-1	1	-1	1	-1	1	-1	1	-1		
B_{2g}	1	-1	1	1	-1	1	1	-1	1	-1	-1	1		
E_{1g}	2	1	-1	-2	0	0	2	1	-1	-2	0	0	(R_x, R_y)	(xz, yz)
E_{2g}	2	-1	-1	2	0	0	2	-1	-1	2	0	0		(x^2-y^2, xy)
A_{1u}	1	1	1	1	1	1	-1	-1	-1	-1	-1	-1		
A_{2u}	1	1	1	1	-1	-1	-1	-1	-1	-1	1	1		z
B_{1u}	1	-1	1	-1	1	-1	-1	1	-1	1	-1	1		
B_{2u}	1	-1	1	-1	-1	1	-1	1	-1	1	1	-1		
E_{1u}	2	1	-1	-2	0	0	-2	-1	1	2	0	0		(x, y)
E_{2u}	2	-1	-1	2	0	0	-2	1	1	-2	0	0		

$D_{\infty h}$ See Table 7.6 p. 163

(h) D_{nd} POINT GROUPS

D_{2d}	I	$2S_4$	C_2	$2C_2'$	$2\sigma_d$		
A_1	1	1	1	1	1		$x^2+y^2,\ z^2$
A_2	1	1	1	-1	-1	R_z	
B_1	1	-1	1	1	-1		x^2-y^2
B_2	1	-1	1	-1	1		$z,\ xy$
E	2	0	-2	0	0	(R_x, R_y)	$(x, y)(xz, yz)$

D_{3d}	I	$2C_3$	$3C_2$	i	$2S_6$	$3\sigma_d$		
A_{1g}	1	1	1	1	1	1		$x^2+y^2,\ z^2$
A_{2g}	1	1	-1	1	1	-1	R_z	
E_g	2	-1	0	2	-1	0	(R_x, R_y)	$(x^2-y^2, xy)\ (xz, yz)$
A_{1u}	1	1	1	-1	-1	-1		
A_{2u}	1	1	-1	-1	-1	1		z
E_u	2	-1	0	-2	1	0		(x, y)

D_{4d}	I	$2S_8$	$2C_4$	$2S_8^3$	C_2	$4C_2'$	$4\sigma_d$		
A_1	1	1	1	1	1	1	1		$x^2+y^2,\ z^2$
A_2	1	1	1	1	1	-1	-1	R_z	
B_1	1	-1	1	-1	1	1	-1		
B_2	1	-1	1	-1	1	-1	1		z
E_1	2	$\sqrt{2}$	0	$-\sqrt{2}$	-2	0	0		(x, y)
E_2	2	0	-2	0	2	0	0		(x^2-y^2, xy)
E_3	$2-\sqrt{2}$		0	$\sqrt{2}$	-2	0	0	(R_x, R_y)	(xz, yz)

(i) CUBIC POINT GROUPS

T_d See Table 7.5 p. 159 O_h See Table 7.3 p. 155

Index

Index of Compounds Discussed in the Main Text